"For anyone who grew   a
or Middle-earth, this ͺ   ... ͺͺ ͺ ͺₒᵧ aₙₒ a revelation—you'll be
reminded just how deep those images went into your heart. I'm
pretty sure the best place to read it is with your back against a tree
trunk on a sunny day—but if it's cold and snowy out, these pages
will summon that summer in your soul."

Bill McKibben, author of *The Flag, the Cross, and the Station Wagon*

"In this timely book, Kristen Page issues a compelling prophetic call
for conservative evangelicals who have often ignored climate crisis
warnings—and also love Lewis and Tolkien—to care for God's good
creation. She shows how losing ourselves within the beauty of Narnia
and Middle-earth's imaginary landscapes can awaken our capacity
for ecological wonder and humility, leading us to extol the glory of
God and respect the *imago Dei* in all of creation. Page and her conversation partners model how scholars in science, theology, and literature might come together to address crucial issues of our time."

Gary S. Selby, professor of ministerial formation at Emmanuel Christian
Seminary and author of *Pursuing an Earthy Spirituality: C. S. Lewis and
Incarnational Faith*

"Kristen Page's exploration of Narnia, Middle-earth, and our own
wondrous earth is sure to inspire readers to return to the tales of
Lewis and Tolkien and see afresh the beauty and brokenness of the
world just outside their front door. Page is an engaging and insightful guide to these landscapes, and her reflections are en-
livened by thoughtful responses from colleagues at Wheaton.
Highly recommended for fans of Tolkien and Lewis, for those who
love literature and ecology, and really for all of us whose capacity
for wonder will be expanded by this delightful little book."

Jonathan A. Moo, professor of New Testament and environmental
studies at Whitworth University

"I am looking forward to a day in which science and the arts will
end their centuries-long separation. In particular, I think the
science of ecology is the best-known living discourse for recovering the ethos of premodern approaches to the cosmos. I read
this book with highlighter in hand, eager to find a way into this
promising dialogue."

Jason M. Baxter, associate professor of fine arts at Wyoming Catholic
College and author of *The Medieval Mind of C. S. Lewis*

"*The Wonders of Creation* is a creative, insightful, and well-written book. It is, furthermore, a timely tome that shows how fictional landscapes, such as those created by C. S. Lewis and J. R. R. Tolkien, can inspire us to care for the damaged landscapes of our world today. Drawing on careful readings of Lewis and Tolkien, ecologist Kristen Page weaves a tapestry of reflections on ecological literacy, lament, and wonder, to which Christina Bieber Lake, Emily Hunter McGowin, and Noah Toly offer short responses. The thoughtful writing in *The Wonders of Creation* will foster our care of our home places."

Steven Bouma-Prediger, Hope College, author of *For the Beauty of the Earth*

THE
WONDERS
OF
CREATION

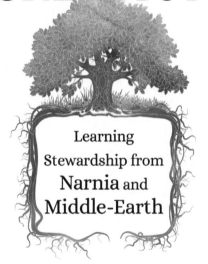

Learning
Stewardship from
Narnia and
Middle-Earth

KRISTEN PAGE

With contributions from Christina Bieber Lake,
Noah J. Toly, and Emily Hunter McGowin

ivp
Academic
An imprint of InterVarsity Press
Downers Grove, Illinois

InterVarsity Press
P.O. Box 1400 | Downers Grove, IL 60515-1426
ivpress.com | email@ivpress.com

InterVarsity Press® is the publishing division of InterVarsity Christian Fellowship/USA®. For more information, visit intervarsity.org.

All Scripture quotations, unless otherwise indicated, are taken from The Holy Bible, New International Version®, NIV®. Copyright © 1973, 1978, 1984, 2011 by Biblica, Inc.™ Used by permission of Zondervan. All rights reserved worldwide. www.zondervan.com. The "NIV" and "New International Version" are trademarks registered in the United States Patent and Trademark Office by Biblica, Inc.™

The publisher cannot verify the accuracy or functionality of website URLs used in this book beyond the date of publication.

Cover design and image composite: David Fassett
Interior design: Jeanna Wiggins

ISBN 978-1-5140-0435-7 (print) | ISBN 978-1-5140-0436-4 (digital)

Printed in the United States of America ∞

Library of Congress Cataloging-in-Publication Data
A catalog record for this book is available from the Library of Congress.

29 28 27 26 25 24 23 22 | 8 7 6 5 4 3 2 1

To all who have journeyed with me

through the many landscapes of my life,

but especially to Dr. Paul Farmer, who taught us

all the importance of navigating

with accompaniment.

CONTENTS

ACKNOWLEDGMENTS

To be honest, this project took me by surprise. I never would have thought to pursue this type of scholarship had it not been for Marjorie Mead. Thank you, Marj, for believing in me and encouraging me to explore how two of my passions, literature and science, work together. My scholarship and teaching will never be the same! I also want to thank Walter and Darlene Hansen for the generous gift that provided this opportunity and for the ways they encourage me through their pursuit of wonder in God's creation. To everyone at the Wade Center, especially Crystal and David Downing, and Laura Schmidt, thank you for helping me navigate this new area of scholarship.

So many friends and colleagues helped me throughout the project. Phoebe Beheler has encyclopedic knowledge of Tolkien, and always challenged me to think about my arguments in new ways. Thanks, Phoebe, for being my sounding board and always being honest about my questionable ideas. I must thank Joanne Ebihara for the cups of tea and engaging conversations that helped me shape my earliest ideas into questions worth pursuing. Many other friends and colleagues walked with me (literally and figuratively) and helped me organize my thoughts. Mary Gonzalez, thank you for wondering with me and showing me how people can form new ideas about creation care through the practice of intentional wonder. Thank you, Laura Yoder, Nadine Rorem, Jovanka Tepavčević, Jim Beitler, Amy Black, Teresa Brown, Amanda

Beheler, and my colleagues in Biological and Health Sciences for your smiles, encouragement, and practical help as I was writing.

My family is amazing, and they are my biggest fans and best supporters. Mom and Dad (Anna Laura and Oscar Page), thanks for listening to me practice my talks a million times and thanks for driving me around to search out the photographs I needed to illustrate my ideas. Rick and Wren, you know you make every day a wonder! Thank you for joining me on my adventures across so many landscapes. I marvel that I am so blessed to have you walking by my side while I explore and wonder at all that God has created. Thank you!

INTRODUCTION

Walter Hansen

I *heard creation's call the summer* before hearing Kristen's lectures with that title. In June 2020, Darlene and I left our condo in downtown Chicago and drove west in our RV to explore national parks. As for much of the country, it was a dark, grim time of pandemic in our city: a stay-at-home order; the closure of restaurants, music venues, and churches; and even our favorite Lake Shore Park paths shut out all but digital access to friends and nature. We longed to be out in the forests and mountains. So, we escaped to Rocky Mountain National Park, Arches National Park, and Grand Teton National Park, our top three favorites of all the parks we have visited. As we were hiking trails through Colorado columbine and forests, climbing breathless over mountain crests, swimming in clear, cold mountain lakes, and viewing innumerable stars over the Arches, we joined our voices with creation's praise of the Creator.

Hearing Kristen's lectures renewed and deepened my love and sense of stewardship for the wonders of God's creation. Her exuberant love and in-depth knowledge of forests made her the perfect guide as she taught me how to "read landscapes" in the real world and in the fictional worlds of Narnia and Middle-earth. I learned from her how to explore the beauty of nature, "like a visiting scarlet tanager in the top of a bur oak or a roaming red fox in the honeysuckle hedge." I gasped with delight when I saw her photos of birds: eastern bluebird, bald eagle, indigo bunting, red-fronted macaw, great horned owl, cardinal, and Nashville warbler. I also gained a fresh understanding of God's creation from her insightful readings of the "subcreations" of Tolkien and Lewis. The exceptional clarity of Kristen's vision led me to see environmental truths told in their stories about the beauty and vulnerability of nature.

Because Kristen's academic research specializes in wildlife diseases and how landscapes are degraded by human use and misuse of natural resources, she speaks forcefully about the destruction of nature caused by human self-indulgence and wastefulness. The shock of the hobbits when they returned to their beloved Shire and saw that it was being destroyed by the same evil they saw in Mordor mirrors Kristen's shock when she worked every day doing research in "urban brownfields." The description of Mordor—"the gasping pools were choked with ash and crawling muds, sickly white and grey"—resonated with her experience in former industrial sites contaminated with chemicals: "The smell was nauseating and every few steps there were piles of what looked like chalky white gooey green chemical residue." Kristen discovered that there are at least one million brownfields in the United States, most of them in urban, low-income areas, exposing 25 percent of children in the United States to toxic waste. Thankfully, most people live in safe homes with sanitation systems that keep us separate from our waste and toxic exposure. Yet except for recreational visits to well-kept parks, we live in a way that largely separates us from nature.

Because of this separation, we urgently need new ways of awakening to our calling to care for creation. Fortunately, as

Kristen so engagingly points out, Lewis and Tolkien offer a powerful means to aid us in that recovery. Through their stories, we are invited to imaginatively experience both the joy of living wisely in God's beautiful world as well as the desolation that occurs when that beauty is lost. As a result, this book can aid all of us in acquiring heightened awareness of our essential role as stewards for the natural world. Indeed, as you read *The Wonders of Creation*, you will find yourself moving from laughing with wonder as you see the portrayal of the beauty of God's creation as reflected in the stories of Narnia and Middle-earth to lamenting the devastation of nature caused by human greed and neglect to listening more deeply to the praise and the pain of creation.

Follow creation's call with an open heart and open hands. Hearts full of love for this beautiful planet will move our hands to care for animals, forests, flowers, birds, rivers, lakes, and oceans where we live. God put us in this garden to keep it. Let's not lose it.

The Ken and Jean Hansen Lectureship

I'm motivated to set up a lectureship in honor of my parents, Ken and Jean Hansen, at the Wade Center primarily because they loved Marion E. Wade. My father began working for Mr. Wade in 1946, the year I was born. He launched my father and mentored him in his business career. Often when I look at the picture of Marion Wade in the Wade Center, I give thanks to God for his beneficial influence in my family and in my life.

After Darlene and I were married in December 1967, the middle of my senior year at Wheaton College, we invited Marion and Lil Wade for dinner in our apartment. I wanted Darlene to get to know the best storyteller I've ever heard.

When Marion Wade passed through death into the Lord's presence on November 28, 1973, his last words to my father were, "Remember Joshua, Ken." As Joshua was the one who followed Moses to lead God's people, my father was the one who followed Marion Wade to lead the ServiceMaster Company.

After members of Marion Wade's family and friends at ServiceMaster set up a memorial fund in honor of Marion Wade at Wheaton College, my parents initiated the renaming of Clyde Kilby's collection of papers and books from the seven British authors—C. S. Lewis, J. R. R. Tolkien, Dorothy L. Sayers, George MacDonald, G. K. Chesterton, Charles Williams, and Owen Barfield—as the Marion E. Wade Collection.

I'm also motivated to name this lectureship after my parents because they loved the literature of these seven authors whose papers are now collected at the Wade Center.

While I was still in college, my father and mother took an evening course on Lewis and Tolkien with Dr. Kilby. The class was limited to nine students so that they could meet in Dr. Kilby's living room. Dr. Kilby's wife, Martha, served tea and cookies.

My parents were avid readers, collectors, and promoters of the books of the seven Wade authors, even hosting a book club in their living room led by Dr. Kilby. When they moved to Santa Barbara in 1977, they named their home Rivendell, after the beautiful house of the elf Lord Elrond, whose home served as a welcome haven to weary travelers as well as a cultural center for Middle-earth history and lore. Family and friends who stayed in their home know that their home fulfilled Tolkien's description of Rivendell:

> And so at last they all came to the Last Homely House, and
> found its doors flung wide. . . . [The] house was perfect

whether you liked food, or sleep, or work, or story-telling, or singing, or just sitting and thinking best, or a pleasant mixture of them all. . . . Their clothes were mended as well as their bruises, their tempers and their hopes. . . . Their plans were improved with the best advice.[1]

Our family treasures many memories of our times at Rivendell, highlighted by storytelling. Our conversations often drew from images of the stories of Lewis, Tolkien, and the other authors. We had our own code language: "That was a terrible Bridge of Khazad-dûm experience." "That meeting felt like the Council of Elrond."

One cold February, Clyde and Martha Kilby escaped the deep freeze of Wheaton to thaw out and recover for two weeks at my parents' Rivendell home in Santa Barbara. As a thank-you note, Clyde Kilby dedicated his book *Images of Salvation in the Fiction of C. S. Lewis* to my parents. When my parents set up our family foundation in 1985, they named the foundation Rivendell Stewards' Trust.

In many ways, they lived in and they lived out the stories of the seven authors. It seems fitting and proper, therefore, to name this lectureship in honor of Ken and Jean Hansen.

Escape for Prisoners

The purpose of the Hansen Lectureship is to provide a way of escape for prisoners. J. R. R. Tolkien wrote about the positive role of escape in literature:

I have claimed that Escape is one of the main functions of fairy-stories, and since I do not disapprove of them, it is plain that I do not accept the tone of scorn or pity with

---

[1] J. R. R. Tolkien, *The Hobbit* (London: Unwin Hyman, 1987), 50-51.

which "Escape" is now so often used: a tone for which
the uses of the word outside literary criticism give no
warrant at all. In what the misusers of Escape are fond
of calling Real Life, Escape is evidently as a rule very
practical, and may even be heroic.[2]

Note that Tolkien is not talking about escapism or an avoidance
of reality, but rather the idea of escape as a means of providing
a new view of reality, the true transcendent reality that is
often screened from our view in this fallen world. He adds:

Evidently we are faced by a misuse of words, and also by
a confusion of thought. Why should a man be scorned, if,
finding himself in prison, he tries to get out and go home?
Or if, when he cannot do so, he thinks and talks about
other topics than jailers and prison-walls? The world
outside has not become less real because the prisoner
cannot see it. In using Escape in this [derogatory] way the
[literary] critics have chosen the wrong word, and, what
is more, they are confusing, not always by sincere error,
the Escape of the Prisoner with the Flight of the Deserter.[3]

I am not proposing that these lectures give us a way to escape
from our responsibilities or ignore the needs of the world
around us, but rather that we explore the stories of the seven
authors to escape from a distorted view of reality, from a
sense of hopelessness, and to awaken us to the true hope of
what God desires for us and promises to do for us.

C. S. Lewis offered a similar vision for the possibility that
such literature could open our eyes to a new reality:

---

[2] J. R. R. Tolkien, "On Fairy-Stories," in *Tales from the Perilous Realm* (Boston:
Houghton Mifflin, 2008), 375.
[3] Tolkien, "On Fairy-Stories," 376.

We want to escape the illusions of perspective ... we
want to see with other eyes, to imagine with other imaginations, to feel with other hearts, as well as with our
own. ... The man who is contented to be only himself,
and therefore less a self, is in prison. My own eyes are
not enough for me, I will see through those of others. ...
In reading great literature I become a thousand men yet
remain myself. ... Here as in worship, in love, in moral
action, and in knowing, I transcend myself; and am never
more myself than when I do.[4]

The purpose of the Hansen Lectureship is to explore the great
literature of the seven Wade authors so that we can escape
from the prison of our self-centeredness and narrow, paro-
chial perspective in order to see with other eyes, feel with
other hearts, and be equipped for practical deeds in real life.

As a result, we will learn new ways to experience and extend
the fulfillment of our Lord's mission: "to proclaim freedom for
the prisoners and recovery of sight for the blind, to set the
oppressed free" (Lk 4:18-19 NIV).

---

[4]C. S. Lewis, An Experiment in Criticism (Cambridge: Cambridge University Press, 1965), 137, 140-41.

STEPPING OUT OF
THE WARDROBE

Searching Fictional Landscapes
to Guide Our View of Our Own World

I have always gravitated toward forests, and I think I've always connected to them deeply. In fact, there is a recording of a conversation between my father and me, as a two-year-old, in which when asked, "Where is heaven?" I exclaimed with certainty, "It's in the woods!" I love every kind of forest—those I physically walk through and those I mentally walk through as I read. As an ecologist I spend a lot of time exploring and learning in real forests, thus I can experience fictional forests more deeply with all of my senses. I have no doubt that I can experience literary landscapes so deeply because of my experiences in physical landscapes. As a result, I wonder if those who spend more time in fictional landscapes than actual ones might start to experience nature differently. Could literary landscapes teach us to see creation in a new way and possibly even motivate readers toward environmental stewardship?

One of my favorite literary landscapes is the place between worlds in C. S. Lewis's *The Magician's Nephew*. When I read about this place, I am transported to the many forested landscapes that I have visited in my life. Here is Lewis's description of Digory's first impressions:

> All the light was green light that came through the leaves: but there must have been a very strong sun overhead, for this green daylight was bright and warm. It was the quietest wood you could possibly imagine. There were no birds, no insects, no animals, and no wind. You could almost feel the trees growing. The pool he had just got out of was not the only pool. There were dozens of others—a pool every few yards as far as his eyes could reach. You could almost feel the trees drinking the water up with their roots. This wood was very much alive. When he tried to describe it afterward Digory always said, "It was a rich place: as rich as plumcake." The strangest thing was that, almost before he had looked about him, Digory had half forgotten how he had come there. . . . If anyone had asked him "Where did you come from?" he would probably have said, "I've always been here." That was what it felt like—as if one had always been in that place and never been bored although nothing had ever happened. As he said long afterward, "It's not the sort of place where things happen. The trees go on growing, that's all."[1]

What did you imagine while reading this? While it is hard for me to imagine literal stillness and quiet ("no birds, no insects, no animals, no wind") in any of the forests I know, I understand what is being described. When I am in a forest, I

---

[1]C. S. Lewis, *The Magician's Nephew* (New York: Collier, 1955), 29.

experience a stillness—a quieting of my soul. Like Lewis, I understand that forests are "not the sort of place(s) where things happen"—at least not the everyday things that distract us from what is important. In forests, I find that I can join the psalmist in experiencing this promise: "Be still, and know that I am God" (Ps 46:10). When I read about Lewis's forest in the place between worlds, I can truly understand and experience this place, perhaps as Digory did in *The Magician's Nephew*, because I have experience in actual forests. In fact, experiencing this fictional landscape makes me want to go and find similar places in real life so that I can experience "trees (that) go on growing." Don't you wish this too?

I love reading as much as I love exploring landscapes (whether forests, mountains, prairies, or wetlands). I read books to learn, to worship, to teach, to escape, and to relax. I am particularly drawn to books about nature or those with settings that enable me to feel part of nature through detailed descriptions. By reading such books, I have gained a desire both to be in nature as well as to protect it. I have also developed what Aldo Leopold, a famous conservation biologist and the father of wildlife biology, defines as a *land ethic*. I understand that I am part of a community that extends beyond people to include plants, animals, and all of creation.[2] Robert Macfarlane, an author of several books about place, literature, and imagination, writes, "Whenever I ask professional conservationists what first inspired them to get involved in the protection of the environment, they invariably mention either a book or a place."[3]

---

[2] Aldo Leopold, *A Sand County Almanac with Essays on Conservation from Round River* (New York: Ballantine, 1966).
[3] Robert Macfarlane, in "Landscape, Literature, Life," World.edu, March 4, 2012, http://world.edu/landscape-literature-life/.

It is my experience that many of us who have pursued conservation as a vocation connect strongly to places, even
places discovered in books. We can easily connect fictional
places with places we know that need protection or resto-
ration. Love for these places is enough to motivate us to environmental action. It is also true that those of us who pursue
environmental stewardship are drawn to writing about
nature. Macfarlane suggests that most people, however, are
not motivated to action simply through reading "nature
writing." He argues that those of us who are motivated by
writings of naturalists are probably already interested and
inclined toward environmental action. To those without a
deep-seated inclination toward or interest in the natural
world, books about nature can seem uninteresting, pious, or
even tending toward propaganda.[4] However, these same
people often are strongly drawn to the grand fictional landscapes created by authors like C. S. Lewis and J. R. R. Tolkien.
For some of these readers, as they learn to care about these
fictional locations, they are also awakened to the beauty of
the created world in which they live. Accordingly, they may
begin to care about the conservation and protection of the
landscapes surrounding them.

In their work *Narnia and the Fields of Arbol*, Dickerson and
O'Hara argue that "people are sometimes willing to listen to
ideas that come in the form of story that they would not listen
to in the form of abstract arguments."[5] Further, Patrick Curry
also maintains that in order for readers to be motivated to environmental action from the stories they read, they must have

---

[4]Robert Macfarlane, "New Words on the Wild," *Nature* 498 (2013): 166-67.
[5]Matthew Dickerson and David O'Hara, *Narnia and the Fields of Arbol: The Environmental Vision of C. S. Lewis* (Lexington: University Press of Kentucky,
2008).

formed a personal connection to these fictional landscapes.[6] J. R. R. Tolkien and C. S. Lewis both created landscapes that readers seem to connect with, including descriptions of fictional places that their readers know intimately and love to explore. For example, Tolkien's "subcreation"[7] of Middle-earth provides a setting for his stories that connects the readers to real landscapes ranging from swamps and forests to mountains and grasslands.[8] In particular, the hobbits' beloved Shire was based on the rural land in the West Midlands of England, where Tolkien spent his boyhood. Likewise, Lewis's creation of Narnia reflects the landscapes he loved to walk across in the Belfast countryside.[9] Thus, both Tolkien's and Lewis's personal experience of the natural world allowed them to create familiar places that readers can recognize and envision for themselves as they read the stories. Stories speak of truths,[10] including environmental truths about the vulnerabilities of nature when actions within the stories destroy rather than protect the created world.

Reading Landscapes

Everyone has some kind of connection to place. When I teach ecology, I start the course asking my students to think of their favorite place in nature and describe it to me. I remind them that there are many aspects to place—there are elements that

---

[6]Patrick Curry, "Tolkien and Nature," Tolkien Estate, 2015, www.tolkienestate .com/writing/patrick-curry-tolkien-and-nature/.

[7]Verlyn Flieger and Douglas A. Anderson, eds., *Tolkien on Fairy-Stories* (London: HarperCollins, 2008).

[8]Dinah Hazell, *The Plants of Middle-earth* (Kent, OH: Kent State University Press, 2006).

[9]Dickerson and O'Hara, *Narnia and the Fields of Arbol.*

[10]Matthew Dickerson and Jonathan Evans, *Ents, Elves, and Eriador: The Environmental Vision of J. R. R. Tolkien* (Lexington: University Press of Kentucky, 2006).

can be described using our senses: what you see, smell, hear, taste; relationships that require a bit of observation and remembering; and also processes like climate that shape each of these elements. Essentially, what the students are doing in this exercise is describing their home landscape. My students are thinking about a familiar place, one they connect with and are drawn to. This can be any place—their actual home or where they find the stillness the Psalmist describes. I call this activity "reading landscapes." It involves scanning, recognizing, identifying, and interpreting a place with the hope of learning, worshiping, teaching, escaping, or relaxing.[11] Reading the landscape involves paying attention to one's surroundings in order to understand what attributes make the place unique. As the landscape is explored, the "reader" considers the common species making up the specific community (like robins in a maple tree and gray squirrels hiding acorns), delights with every surprising discovery (like a visiting scarlet tanager in the top of a bur oak or a roaming red fox in the honeysuckle hedge), and considers how they personally interact with the place.

Reading landscapes includes not only the flora and fauna but also the habitat. Interestingly, landscape is a term used by many disciplines to describe the attributes of a place.[12] For example, scientists might be interested in the elements of the land that explain natural processes, like ecosystem services,[13] while artists may be interested in connecting feelings or

[11]David Schur, An Introduction to Close Reading (Cambridge, MA: Harvard University Press, 1998); Stanislas Dehaene, Reading in the Brain: The Science and Evolution of a Human Invention (New York: Viking, 2009).

[12]Thomas Kirchoff, Ludwig Trepl, and V. Vicenzotti, "What Is Landscape Ecology? An Analysis and Evaluation of Six Different Conceptions," Landscape Research 38 (2013): 33-51.

[13]Kirchoff, Trepl, Vicenzotti, "What Is Landscape Ecology?"

memories or longings to the land by focusing on elements of the landscape that are familiar. This is why we are not disoriented when we enter Middle-earth at the beginning of *The Hobbit*, for what child has not dreamed of digging out a fort or a hideaway? However, I'm guessing most of these attempts result in a hole that isn't nearly as welcoming as the child envisioned. This is where Tolkien starts with the development of his landscape: "In a hole in the ground there lived a hobbit. Not a nasty, dirty, wet hole, filled with the ends of worms and an oozy smell, nor yet a dry, bare, sandy hole with nothing in it to sit down on or to eat: it was a hobbit-hole, and that means comfort."[14]

The description of what the hole is not, and the similarity to the shared experiences of many children, is exactly what orients the reader to the "comfort" of a hobbit-hole. Shared experiences allow us to understand landscapes, whether we have spent time in them or not. In this way, the landscapes of Tolkien and Lewis extend beyond mere settings for a story; these authors have created worlds with specific geologies, geographies, ecologies, and cultures. Just like in actual landscapes, readers can experience these created landscapes at different scales. For example, in the terrain of *The Hobbit*, the reader can experience the detail of the path:

> As they went on Bilbo looked from side to side for something to eat; but the blackberries were still only in flower, and of course there were no nuts, not even hawthorn-berries. . . . They still went on and on. The rough path disappeared. The bushes, and the long grasses between the boulders, the patches of rabbit-cropped turf, the

---

[14]J. R. R. Tolkien, *The Hobbit, or There and Back Again* (London: HarperCollins, 1991), 3.

thyme and the sage and the marjoram, and the yellow rockroses all vanished, and they found themselves at the top of a wide steep slope of fallen stones, the remains of a landslide.[15]

Readers can also experience the grand, sweeping vistas of the journey: "Now they could look back over the lands they had left, laid out behind them far below. Far, far away in the West, where things were blue and faint, Bilbo knew their lay his own country of safe and comfortable things, and his little hobbit-hole."[16]

Regardless of the scale, there are elements of the landscape that orient the reader to a place of comfort because they connect the reader to elements that are familiar. The created landscapes of Lewis and Tolkien are believable because each author spent time in nature learning to read actual landscapes in addition to spending time in literary landscapes as they read throughout their lives.

Imagined Landscapes

C. S. Lewis created imagined landscapes for most of his life. In *Surprised by Joy*, he describes his boyhood fictional creation "Animal Land" as a place of "anthropomorphized beasts" that entertained his imagination. However, he describes his drawings as lacking beauty and having a "shocking ignorance of natural form." Lewis says, "This absence of beauty, now that I come to think of it, is characteristic of our childhood."[17] However, a "biscuit-tin garden" made by his brother with

---

[15]Tolkien, *The Hobbit*, 113.
[16]Tolkien, *The Hobbit*, 65.
[17]C. S. Lewis, *Surprised by Joy: The Shape of My Early Life* (New York: Harper-One, 1955), 5.

moss, twigs, and flowers was brought to the nursery one day, and "that was the first beauty [he] knew."[18] The moss-garden of Lewis's childhood, perhaps, was the seed for the imagined landscapes of his fiction, since he wrote that it first made him aware of nature.[19]

Lewis loved walking across landscapes and was certainly influenced by what he experienced. For example, in October 1918, he wrote in a letter to his boyhood friend Arthur Greeves,

"Savernake Woods," doesn't that breathe of romance? . . . I have been in Savernake Woods this morning. You get clear of the village, cross a couple of fields and then a sunken chalky road leads you right into the wood. It is full of beech and oak but also of those little bushy things that grow out of the earth in four or five different trunks—vide Rackham's woodland scenes in the "Siegfried" illustrations. In places, too, there has been a good deal of cutting down: some people think this spoils a wood but I find it delightful to come out of the thickets suddenly to a half bare patch full of stumps and stacks of piled wood with the sun glinting thro' the survivors. Green walks of grass with thick wood on either side led off the road and we followed one of these down and found our way back by long détours.[20]

As he explains in another letter to his brother, Warren, his experience went beyond "lines and colours, but (extends to) smells, sounds, and tastes."[21] Lewis further describes

---

[18]Lewis, *Surprised by Joy*, 6.
[19]Lewis, *Surprised by Joy*, 6.
[20]Walter Hooper, ed., *They Stand Together: The Letters of C. S. Lewis to Arthur Greeves (1914–1963)* (London: Collins, 1979), 233.
[21]C. S. Lewis, *The Collected Letters*, Vol. 1 (London: HarperCollins, 2000), 558.

how as he walked, he thought of literature or opera and considered how beloved scenes might play out in the landscape where he was walking: "I was always involuntarily looking for scenes that might belong to the Wagnerian world, here a steep hillside covered with firs where Mime might meet Sieglinde, there a sunny glade where Siegfried might listen to the bird . . . But soon . . . nature ceased to be a mere reminder of the books, became herself the medium of the real joy."[22]

Spending time in nature certainly impacted Lewis's creativity, and as a result, his readers experience many of his fictional landscapes in the same way that he explored actual landscapes; as the characters walk or journey, the landscapes unfold. Consider, for example, this passage from *Out of the Silent Planet*, the first book in Lewis's Space Trilogy:

> They walked forward—beside the channel. In a few minutes Ransom saw a new landscape. The channel was not only a shallow but a rapid—the first, indeed, of a series of rapids by which the water descended steeply for the next half-mile. The ground fell away before them and the canyon—or handramit—continued at a very much lower level. Its walls, however, did not sink with it, and from his present position Ransom got a clearer notion of the lie of the land.[23]

Because Lewis spent much time walking across landscapes, he is able to create something beautiful and familiar in an entirely new landscape on a planet previously unknown to the reader.

---

[22]Lewis, *Surprised by Joy*, 77.
[23]C. S. Lewis, *Out of the Silent Planet* (New York: Scribner, 1996), 63.

Just as Lewis learned to love nature from the biscuit-tin garden created by his brother Warren, Tolkien was introduced to the beauty of the natural world by his mother and her love of botany and gardening.[24] He cultivated his own love of plants, especially trees, throughout his life.[25] He spent enough time in nature reading landscapes to be able to know the intricacies of natural rhythms. In a letter to his son, Christopher, he describes in detail the flowering and leaf-out sequence of the trees: "The oaks were among the earliest trees to be leafed equaling or beating birch, beech and lime etc. Great cauliflowers of brilliant yellow-ochre tasseled with flowers, while the ashes (in the same situations) were dark, dead, with hardly even a visible sticky bud."[26]

Tolkien's observations and knowledge of nature allowed him to create fictional landscapes described in such detail that plant guides have been written and taxonomic keys developed to help the reader navigate the flora of Middle-earth.[27] These familiar elements of landscape allow the reader to become immersed in imaginary worlds. Tolkien explains in a 1964 interview with the BBC that elements of his fictional landscapes were based on the place where "memory and imagination" come together, and that "the Shire is very much like the kind of world in which I first became aware of things."[28] Middle-earth represents the "actual Old world of this planet,"[29]

---

[24]Hazell, *Plants of Middle-earth.*

[25]Humphrey Carpenter and Christopher Tolkien, *The Letters of J. R. R. Tolkien* (Boston: Houghton Mifflin, 2000), 220.

[26]Carpenter and Tolkien, *Letters of J. R. R. Tolkien,* 408.

[27]Hazell, *Plants of Middle-earth*; Walter Judd and Graham Judd, *Flora of Middle-earth: Plants of J. R. R. Tolkien's Legendarium* (New York: Oxford University Press, 2017).

[28]J. R. R. Tolkien, 1964 interview by Denys Geroult, British Broadcasting Corporation, broadcast in 1971.

[29]Carpenter and Tolkien, *Letters of J. R. R. Tolkien,* 220.

but Tolkien does not claim to "relate the shape of the mountains and land-masses";[30] rather he describes his created landscape as "at a different stage of imagination."[31] He explains, "Mine is not an 'imaginary' world, but an imaginary historical moment on 'Middle-earth'—which is our habituation."[32] Because Middle-earth "is our habituation,"[33] we might expect to recognize elements of the "habitable lands of our world."[34]

Landscapes That Transform

Time spent in nature "reading landscapes" helped Lewis and Tolkien create realistic and relatable places for their fictional works to develop, yet these places are much more than settings for stories. Lewis and Tolkien have created worlds that are familiar to their readers: ecosystems that we can understand and even experience as we journey with the sons of Adam, daughters of Eve, or the Fellowship.[35] These created landscapes play a role in transforming characters as they pass on their journeys, and readers might also experience transformations as they participate in the journeys of the characters. For example, landscapes dominated by forests often are important to the action of the story, as if they too are characters. In Lewis's *The Last Battle*, the trees call King Tirian to action (even if indirectly). He is sitting under a "great oak" enjoying "pleasant spring weather" and starts receiving messages of Aslan's return to Narnia from the birds and the squirrels. Just as Tirian is questioning the rumors of the return of Aslan to Narnia, a dryad of a beech tree comes to him with an urgent message.

[30]Carpenter and Tolkien, *Letters of J. R. R. Tolkien*, 220.
[31]Tolkien, 1964 interview by Geroult.
[32]Carpenter and Tolkien, *Letters of J. R. R. Tolkien*, 244.
[33]Carpenter and Tolkien, *Letters of J. R. R. Tolkien*, 244.
[34]Carpenter and Tolkien, *Letters of J. R. R. Tolkien*, 376.
[35]Judd and Judd, *Flora of Middle Earth*.

"Woe, woe, woe!" called the voice. "Woe for my brothers and sisters! Woe for the holy trees! The woods are laid waste. The axe is loosed against us. We are being felled. Great trees are falling, falling, falling."[36]

King Tirian responds by rushing to the aid of the ancient forest, yet he is too late for many of the trees. "Right through the middle of that ancient forest—that forest where the trees of gold and of silver had once grown . . . a broad lane had already been opened. It was a hideous lane like a raw gash in the land, full of muddy ruts where felled trees had been dragged down to the river."[37]

Tirian's response to the injustices he discovers should provide an example for us. There are many real landscapes that are experiencing similar destruction, yet we do not respond. Perhaps our response to the devastating scene in *The Last Battle* is more compassionate than our response to the devastation of real landscapes because in the story we understand the trees to be "alive" and able to speak. In the short space of a chapter, we are able to realize that these trees are being devastated for the economic gain of those who do not even live in the country—the Calormens. Do we respond differently when we can directly link devastation in our world to consumerism and economic pursuit? The same story is playing out in our lives, too, yet because we are unaware or unconcerned, we fail to act. Actual forests are being lost at a rate of 200,000 km^2/year,[38] and just as in the story, the trees are being sold to people who do not live among the forests. The forests of Brazil, the Democratic Republic of Congo, and

---

[36]C. S. Lewis, *The Last Battle* (New York: Collier, 1956), 16.
[37]Lewis, *Last Battle*, 16.
[38]Mauro Bologna and Gerardo Aquino, "Deforestation and World Population Sustainability: A Quantitative Analysis," *Scientific Reports* 10 (2020): 7631.

Indonesia are being cut at the fastest rates in the world to make way for beef production, soy production, and palm oil plantations to satisfy the consumers of the developed world.[39] Could the compassion for the forest stirred by Lewis transform our reactions to the devastation of real forests?

Tolkien is known for his love of trees, and like Lewis he created landscapes that transformed characters and readers alike. In a letter to the editor of the *Daily Telegraph*, Tolkien describes the importance of trees and forests in his works and how the actions of those entering forests changed the character of the forest.

> In all my works I take the part of trees as against all their enemies. Lothlórien is beautiful because there the trees were loved; elsewhere forests are represented as awakening to consciousness of themselves. The Old Forest was hostile to two legged creatures because of the memory of many injuries. Fangorn Forest was old and beautiful, but at the time of the story tense with hostility because it was threatened by a machine-loving enemy. Mirkwood had fallen under the domination of a Power that hated all living things but was restored to beauty and became Greenwood the Great before the end of the story.[40]

Tolkien's landscapes, especially forests, are alive and make important contributions to the story. Rebecca Merkelbach

---

[39]"Earth's Forests Are Being Cut Down, and They Are Being Cut Down Fast," The World Counts (2020), www.theworldcounts.com/challenges/planet-earth /forests-and-deserts/rate-of-deforestation; "What Are the Biggest Drivers of Tropical Deforestation? They May Not Be What You Think," *World Wildlife Magazine* (Summer 2018), www.worldwildlife.org/magazine/issues/summer -2018/articles/what-are-the-biggest-drivers-of-tropical-deforestation.

[40]Carpenter and Tolkien, *Letters of J. R. R. Tolkien*, 419.

describes Tolkien's forests as "transformative spaces," and just as the forest is characterized by its history, those entering the forest are changed when they leave.[41] The hobbits are reluctant to enter the Old Forest. They recognize the forest as "alive . . . (and) more aware of what is going on . . . than things (were) in the Shire." They comment that "the trees do not like strangers" and often "bar their way" by shifting.[42] As they leave the darkness of the forest behind and begin to "have some notion of where (they were)" along the River Withywindle, they encounter Old Man Willow. Tom Bombadil arrives just as Old Man Willow has trapped Merry and Pippin—perhaps exacting justice for past wrongs against the forest. The hobbits' experience in the Old Forest transforms them and prepares them to continue their journey, but their experience with Tom Bombadil also teaches them something about their place in nature, and perhaps even their responsibility toward it: "As they listened, they began to understand the lives of the Forest, apart from themselves, indeed to feel themselves as the strangers where all other things were at home."[43]

Treebeard furthers the education of Merry and Pippin about how landscapes reveal the ambitions of those living there. Fangorn forest was once vast and full of thick, strong, young trees, but at the time of this meeting, it was a remnant of its former majesty. It had succumbed to the desires of Saruman the wizard, who at one point seemed interested in learning and knowing about the forest but

---

[41]Rebecca Merkelbach, "Deeper and Deeper into the Woods: Forests as Places of Transformation in *The Lord of the Rings*," in J. R. R. *Tolkien: The Forest, and the City*, ed. Helen Conrad-O'Brian and Gerald Hynes (Portland: Four Courts, 2013), 57-66.

[42]J. R. R. Tolkien, *The Lord of the Rings: The Fellowship of the Ring* (London: HarperCollins, 2012), 110-11.

[43]Tolkien, *The Fellowship of the Ring*, 129-30.

now does not care for "growing things, except as far as they serve him for the moment."[44] So when we enter Fangorn forest as readers, we experience deforestation. The orcs and Saruman are "making havoc" cutting trees to "feed the fires of Orthanc" but also taking more than was needed and wasting many trees by leaving them to rot.[45] Treebeard laments the way that knowledge of the forest has allowed Saruman to extract resources to fuel the fires that will help him to attain more power. In his lament, Treebeard challenges us to action!

> Curse him, root and branch! Many of those trees were my friends, creatures I had known from nut and acorn; many had voices of their own that are lost for ever now. And there are wastes of stump and bramble where once there were singing groves. I have been idle. I have let things slip. It must stop! . . . "I will stop it!" he boomed. "And you shall come with me. You may be able to help me."[46]

Now, when I walk through forests and trip over a root, I wonder if it was my clumsiness or the forest's intention. I no longer experience forests the way that I did before I spent time journeying through the literary forests of Lewis and Tolkien. Is it possible that spending time in fictional landscapes might transform all our attitudes about nature?

Consider Our Christian Responsibility

Despite the biblical support for creation care, many Christians believe that environmentalism should be separate from

---

[44]Tolkien, *The Fellowship of the Ring*, 473.
[45]Tolkien, *The Fellowship of the Ring*, 474.
[46]Tolkien, *The Fellowship of the Ring*, 474.

a life of faith. Politicization of environmental issues has resulted in apathy at best and vehement denial of environmental
problems, like climate change, at worst. I have been told by
pastors and theologians that creation care is not urgent and
is at most a second-tier problem for the church. Comments
like these lead many environmentalists to blame Christians
for the current situation of environmental degradation. In the
now famous essay "The Historical Roots of Our Ecological
Crisis," Lynne White asserts that the state of the environment
is rooted in Christocentric attitudes:

> Our science and technology have grown out of Christian
> attitudes toward man's relation to nature which are
> almost universally held not only by Christians and neo-
> Christians but also by those who fondly regard them-
> selves as post-Christians. Despite Copernicus, all the
> cosmos rotates around our little globe. Despite Darwin,
> we are not, in our hearts, part of the natural process. We
> are superior to nature, contemptuous of it, willing to use
> it for our slightest whim.[47]

Aldo Leopold, a forester, conservationist, and influential
writer, would seemingly agree with White. In his famous collection of essays, *A Sand County Almanac*, he encourages the
development of a land ethic that would move people beyond
this Christocentric idea that humans are distinct from nature.
Leopold argued that ethics needs to move beyond individual
relationship (including relationship with God, as in the case
of the Mosaic Decalogue) and social relationships (e.g., the
Golden Rule) to include relationships with the land, animals,

---

[47]Lynn White, "The Historical Roots of Our Ecological Crisis," *Science* 155 (1967):
1206.

and plants.[48] Further, he maintains, when Christians understand Genesis to mean that humans are to have rights to creation through dominion, then "the land-relation (will be)
strictly economic, entailing privileges but not obligations."[49]
This attitude promotes consumerism and excess use of resources, which facilitates the rapid transformation of landscapes and subsequent loss of biodiversity. Lynne White con-
tinues to explain the destructive influence of this attitude:

> The newly elected Governor of California [Ronald
> Reagan], like myself a churchman but less troubled than
> I, spoke for the Christian tradition when he said (as is
> alleged), "When you've seen one redwood tree, you've
> seen them all." To a Christian a tree can be no more than
> a physical fact. The whole concept of the sacred grove is
> alien to Christianity and to the ethos of the West. For
> nearly 2 millennia Christian missionaries have been
> chopping down sacred groves, which are idolatrous be-
> cause they assume spirit in nature. What we do about
> ecology depends on our ideas of the man-nature rela-
> tionship. More science and more technology are not
> going to get us out of the present ecologic crisis until we
> find a new religion, or rethink our old one.[50]

Rethinking our Christian responsibility to steward creation
is especially urgent today as we experience unprecedented
losses of biodiversity and rapidly changing climate. Certainly,
the fictional writings of Lewis and Tolkien were not specifically intended to motivate people toward environmental

---

[48]Leopold, *Sand County Almanac.*
[49]Leopold, *Sand County Almanac*, 238.
[50]White, "The Historical Roots of Our Ecological Crisis," 1206.

action, yet they both acknowledge that their writing changes the reader, just as they had been changed by the writings of others. For example, Lewis realized that he had personally been influenced by literature. He described the effect that George MacDonald's *Phantastes* had on his own life and imagination: "The quality which had enchanted me in his imaginative works turned out to be the quality of the real universe, the divine, magical, terrifying and ecstatic reality in which we all live."[51]

According to Dickerson and O'Hara, Lewis hoped that his stories would impact his readers as he had been influenced by the stories of others. Readers would understand their situations from new perspectives, separate from current politics and selfish desires.[52] Dickerson and Evans argue that Tolkien used myth for storytelling because it fosters imagination, which is critical to developing solutions to societal issues such as environmental problems.[53] Lewis and Tolkien both demonstrate a well-developed land ethic in their writings and seem to be encouraging the reader toward some type of creation care. In fact, some readers spending time with their stories and in the fictional landscapes they have created are changed. This is the case for Bill McKibben, environmental activist, author, and founder of 350 (www.350.org), a global movement against climate change:

> Lewis was entirely important to me—the Narnia books are the most important literature of my life, I think. No "landscape" is more set in my mind than the edge of the

---

[51]C. S. Lewis, *George MacDonald: An Anthology—365 Readings* (London: Centenary, 1946), 26.
[52]Dickerson and O'Hara, *Narnia and the Fields of Arbol*.
[53]Dickerson and O'Hara, *Narnia and the Fields of Arbol*.

sea, with Reepicheep climbing into his tiny canoe. Except perhaps the heaven described in *The Last Battle*—further up and farther in. I've spent my life as an activist, I think, in large part because I spent my boyhood reading C. S. Lewis.[54]

For McKibben, a childhood spent in the fictional landscapes of Narnia cultivated a strong land ethic and sense of justice. The work of his organization reflects many of the ideas Lewis and Tolkien cultivate in their readers, such as creativity, justice, trust, and care in approaching the environment. This is how God calls us to live, and in this way, creation care is an extension of our call to love our neighbor.

Close Reading Develops a Christian Land Ethic

How does time spent in the fictional creations of Lewis and Tolkien transform us? Could reading their fiction really help us cultivate a stronger land ethic? I would argue that the stories of Lewis and Tolkien encourage us to take time for a closer reading of creation so that we can know more about the natural world. This knowledge will transform us so that we can recognize our place as part of creation and our role as stewards. Finally, we change the way we interact with the natural world by cultivating humility in our use and protection of resources, being content with using only what is needed and delighting in creation.

How do we begin to work toward a closer reading of nature? Literature is an important vehicle to introduce children to the natural world. Many believe that literature written for children

---

[54]Bill McKibben, correspondence with the author, April 20, 2020. Used with permission. For more information, see www.350.org.

provides a way to connect understanding and action.[55] In fact, literature might be a critical means toward transforming our actions toward creation because it is becoming less common for children to spend their childhoods in imaginative play outdoors. Less time spent in nature could have serious implications for creation care; separation is introduced, and we could reinforce ideas of dominating and consuming nature. We see this manifest in a phenomenon called "plant blindness," which is described as the inability to recognize or notice plants in our environment, or the inability to recognize the importance of plants and plant biodiversity.[56] Children often are not introduced to plants in school, and much modern children's literature and television emphasize animals over plants.[57] However, plant blindness can be addressed by intentionally introducing children to the natural world via mentors who teach them about botany, as Tolkien's mother did, or through exposure to plants in carefully selected literature.[58] The campaigns to increase children's literacy regarding plants and the environment[59] seem to be falling short. In 2007, the *Oxford Junior Dictionary* removed close to forty words relating to the natural world in order to make space for words

---

[55]Evelyn Freeman, Barbara Lehman, Lilia Ratcheva-Stratieva, and Patricia Sharer, "To the Reader: Special Issue: Sense of Place in Children's Literature," *Bookbird: A Journal of International Children's Literature* 39, no. 3 (2001).

[56]Allen William, "Plant Blindness," *BioScience* 53 (2003): 926; James Wandersee and Elisabeth Schussler, "Toward a Theory of Plant Blindness," *Plant Science Bulletin* 47 (2001): 2–9.

[57]Dafna Lemish and Colleen Russo Johnson, "The Landscape of Children's Television in the US and Canada," The Center for Scholars and Storytellers, UCLA, Rutgers University, and Ryerson University, April 2019; "An Updated Look at Diversity in Children's Books," *School Library Journal* (2019), www.slj.com/?detailStory=an-updated-look-at-diversity-in-childrens-books.

[58]Wandersee and Schussler, "Plant Blindness," 2–9.

[59]James Wandersee and Elisabeth Schussler, "Preventing Plant Blindness," *The American Biology Teacher* 61 (1999): 82–86.

describing our more solitary and digital world.[60] Words like *acorn, wren, bramble, dandelion,* and *willow* were removed to make way for words such as *blog, broadband, chatroom,* and MP3.[61]

If a shift in vocabulary is an indication of current priorities, it is no surprise that fewer people are interested in caring for the environment. It is difficult to care for something you do not know much about or have never experienced. In 2005, with the first edition of *Last Child in the Woods,* Richard Louv sounded the alarm that a childhood without nature could result in many negative effects, including health consequences.[62] Louv coined this effect *Nature-Deficit Disorder,* and though it is not a medical diagnosis (nor was it ever intended to be), there is a significant body of evidence that suggests exposure to screens, which tends to characterize a life indoors, is associated with health conditions among children including obesity, high blood pressure, insulin resistance, depression, suicidal ideation, attention deficit hyperactivity disorder, and dependence behavior.[63] In addition to human health impacts, the trend of spending less time in nature seriously threatens biodiversity and conservation efforts.[64] Louv warns that

---

[60]Robert Macfarlane and Jackie Morris, *The Lost Words: A Spell Book* (Toronto: Anansi, 2017); Alison Flood, "Oxford Junior Dictionary's Replacement of 'Natural' Words with Twenty-first-century Terms Sparks Outcry," *The Guardian,* January 13, 2015, www.theguardian.com/books/2015/jan/13/oxford-junior -dictionary-replacement-natural-words.

[61]Flood, "Oxford Junior Dictionary's Replacement"; Stefan Fatsis, "Panic at the Dictionary," *The New Yorker,* January 30, 2015, www.newyorker.com/books /page-turner/panic-dictionary.

[62]Richard Louv, *Last Child in the Woods: Saving Our Children from Nature-Deficit Disorder* (Chapel Hill: Algonquin, 2008).

[63]Gadi Lissak, "Adverse Physiological and Psychological Effects of Screen Time on Children and Adolescents: Literature Review and Case Study," *Environmental Research* 164 (2018): 149-57.

[64]Andrew Balmford, Lizzie Clegg, Tim Coulson, and Jennie Taylor, "Why Conservationists Should Heed Pokémon," *Science* 295 (2002): 2367.

"increasingly, nature is something to watch, to consume, to wear—to ignore."[65] Children that learn to "consume" nature are in danger of the effects of overconsumption and will grow to be adults still seeking fulfillment from consumption rather than living a life "captive to joy."[66] As Edmund discovered with his desire for Turkish Delight in C. S. Lewis's *The Lion, the Witch and the Wardrobe*, the more he consumed, the more he desired—just to recapture that initial pleasure.[67] When nature is seen as something to consume, we also fail to understand our place and role. This is especially problematic when it comes to making an argument for Christians to respond to the call to care for creation.

Conservative evangelicals are the least likely group in the United States to regard environmental protection as necessary or even biblical.[68] An important reason for this is that many Christians understand that our role as image bearers (*imago Dei*) separates us from creation—many forget that we too have a creation story! Certainly, we have a distinct role in creation; however, when we think of ourselves as being outside of creation, we understand the term *dominion* or *subdue* as permission or even a command to control or consume creation without limits. In fact, when we see ourselves as separate from creation, it is easier to disregard our impacts on natural systems. We begin to think of the world we live in as simply a pretty backdrop for our important task of sharing the gospel. Perhaps we are experiencing a new kind

---

[65]Louv, *Last Child in the Woods*, 2.

[66]Philippians 4:4-5: "Rejoice in the Lord always. I will say it again: Rejoice! Let your gentleness be evident to all. The Lord is near."

[67]Gary Selby, *Pursuing an Earthy Spirituality: C. S. Lewis and Incarnational Faith* (Downers Grove, IL: InterVarsity Press, 2019).

[68]Bernard Zaleha and Andrew Szasz, "Why Conservative Christians Don't Believe in Climate Change," *Bulletin of the Atomic Scientists* 71 (2015): 19-30.

of "plant blindness"—a "creation blindness" exacerbated by a broken understanding of our role in creation. In Genesis, God tasks Adam with keeping the garden (Gen 2:15). The language used for this task is the same language used later in the Old Testament to describe serving in the temple.[69] Thus Adam was tasked with caring for (service to) the place where God dwells. Additionally, Christians acknowledge the incarnation of God in Christ. We worship God, who came and lived among his creation and was part of creation. Modern Christians too often may not consider the incarnation as pointing us back to the idea of creation as the temple where God dwells. Rather when we read Colossians 1, we fail to recognize that when Christ returns to "reconcile all things," creation is included. Instead, our understanding of the gospel is one in which salvation is internal and about an individual's soul. Accordingly, salvation is private, not shared with community, and depends on the belief and acceptance of each individual.[70] As William L. Portier explains, "Modernity has created a situation in which most people do not expect to encounter God in their daily affairs. God is not expected to appear in our shared public life. If we are interested in God, we pray privately or go to church."[71]

Ultimately, when we forget that creation is "the place where God dwells," we forget our own place in creation and begin to think of our "role" in creation as simply utilitarian. If we can transform that mindset and truly begin to see ourselves as part of creation, we will make decisions based on a more expanded view of community. In this way, we actually

---

[69]John Walton, *The Lost World of Genesis One: Ancient Cosmology and the Origins Debate* (Downers Grove, IL: InterVarsity Press, 2010).
[70]William L. Portier, *Tradition and Incarnation: Foundations of Christian Theology* (New York: Paulist Press, 1993).
[71]Portier, *Tradition and Incarnation*, 51.

can begin to participate in the reconciliation that Christ promises:

> The Son is the image of the invisible God, the firstborn over all creation. For in him all things were created: things in heaven and on earth, visible and invisible, whether thrones or powers or rulers or authorities; all things have been created through him and for him. He is before all things, and in him all things hold together. And he is the head of the body, the church; he is the beginning and the firstborn from among the dead, so that in everything he might have the supremacy. For God was pleased to have all his fullness dwell in him, and through him to reconcile to himself all things, whether things on earth or things in heaven, by making peace through his blood, shed on the cross. (Col 1:15-20)

In the fictional writings of Lewis and Tolkien, we can see the importance of such a transformation in understanding. C. S. Lewis personally experienced such a transformation when he became a Christian. He had long pursued joy— "nearly all that I loved I believed to be imaginary; nearly all that I believed to be real I thought grim and meaningless. The exceptions were certain people (whom I loved and believed to be real) and nature herself."[72] After Lewis became a Christian, he began to recognize these joys as opportunities to experience the presence of God: "At best, our faith and reason will tell us that He is adorable, but we shall not have found Him so, not have 'tasted and seen.' Any patch of sunlight in a wood will show you something about the sun which you could never get from reading books on astronomy. These

---

[72]Lewis, *Surprised by Joy.*

pure and spontaneous pleasures are 'patches of Godlight' in the woods of our experience."[73]

Recognizing these opportunities to enjoy the presence of God in creation is an essential first step toward understanding our role as caretakers of the place where God dwells (creation). Many of Lewis's stories describe characters that go through similar transformations. In *The Magician's Nephew*, Digory is motivated by his personal desire to save his mother from an illness. His focus remains on this goal even after experiencing the magnificence of Aslan's creation song. Aslan tasks him with recovering an apple that will protect Narnia from the evil witch. When he arrives at the tree, he is further tempted by the witch to take the apple for himself, as it will surely cure his mother. Digory's choice to reject the witch's tempting taunts and to return the apple to Aslan is a choice for all of creation. Dickerson and O'Hara explain, "The most beneficial thing one can do for oneself is not try to gain dominion over all creation but to acknowledge one's place in it. By caring for the Tree, the Tree cares for Narnia; by doing the will of Aslan, Digory receives his greatest desire."[74] Just as Digory's choice to plant a tree benefited a greater community and protected all of Narnia, our choices about how we use natural resources can impact our neighbors. Our choices to use resources wisely and not to seek pleasure by overconsuming will result in healthier ecosystems that will sustain the nutritional and resource needs of many neighbors, both those from far away and those who live nearby.

Tolkien's fiction can also point us to the importance of a transformation of ideas about our role as created

---

[73]C. S. Lewis, *Letters to Malcolm: Chiefly on Prayer* (London: Geoffrey Bles, 1964), 119-20.

[74]Dickerson and O'Hara, *Narnia and the Fields of Arbol*, Kindle loc. 1644-46.

caretakers. All of the races present throughout Middle-earth are part of creation, even if there is a hierarchy of roles. As we, humans, are made in the image of God to have a special role with the Creator and to be stewards of creation, the Elves seemed to be set apart as they play an important role in preserving beauty and wilderness.[75] Maintaining the beauty of creation perpetuates the praise song of creation! In Middle-earth, there are other races that teach us about stewardship. The Ent(s), "the earthborn, old as mountains,"[76] teach us the importance of wild places and represent the preservation model of stewardship. They provide us with the memory of the land. Treebeard explains to Merry and Pippin the implications of that history—what the land was like untouched: "Those were the broad days! Time was when I could walk and sing all day and hear no more than the echo of my own voice in the hollow hills. The woods were . . . thicker, stronger, younger. And the smell of the air! I used to spend a week just breathing!"[77] What the land became: "We crossed over Anduin and came to their land; but we found a desert: it was all burned and uprooted, for war had passed over it."[78] His thoughts about the future: "I do not like worrying about the future. I am not altogether on anybody's side, if you understand me: nobody cares for the woods as I care for them, not even the Elves nowdays."[79] And finally, Treebeard describes the cost of power and consumption: "[Saruman] is plotting to become a Power. He has a mind of metal and wheels; and

---

[75]Dickerson and Evans, *Ents, Elves, and Eriador.*
[76]J. R. R. Tolkien, *The Two Towers: Being the Second Part of The Lord of the Rings* (London: HarperCollins, 1954), 464.
[77]Tolkien, *Two Towers,* 469.
[78]Tolkien, *Two Towers,* 476.
[79]Tolkien, *Two Towers,* 472.

he does not care for growing things, except as far as they serve him for the moment."[80]

The hobbits represent a sustainable use model of stewardship. They live in an agrarian society and depend on small-scale agriculture. The simple life of hobbits demonstrates a low-impact approach to resource use that is both hospitable to the land (only taking what is needed) and results in a community built on hospitality to neighbors. When ideas about resources slowly change while Frodo is away, we see firsthand the devastating effects on the land and on the community. Frodo's response to the scouring of the Shire is to lead a conservation/stewardship response. Perhaps the most important way Tolkien works to transform our mindset away from thinking that we are "apart from creation" is by teaching us through Frodo and Sam that sometimes we must sacrifice our own desires so that others will have what they need in the future:

> "But," said Sam, and tears started in his eyes, "I thought you were going to enjoy the Shire, too, for years and years, after all you have done." "So I thought too, once. But I have been too deeply hurt, Sam. I tried to save the Shire, and it has been saved, but not for me. It must often be so, Sam, when things are in danger: someone has to give them up, lose them, so that others may keep them. But you are my heir: all that I had and might have had I leave to you."[81]

Like Frodo, are we able to make courageous decisions that will positively impact the future of our world? What kind of environment will we leave behind for future generations?

---

[80]Tolkien, *Two Towers*, 473.

[81]J. R. R. Tolkien, *The Return of the King: Being the Third Part of The Lord of the Rings* (London: HarperCollins, 2002), 1029.

When we spend more time in nature and experience it more often through literature, Lewis and Tolkien would have us understand that we will learn to love God's creation more as we move away from ideas that nature is simply to be consumed. This shift in attitudes will make a significant difference in our choices and actions as we truly begin to apprehend that we are *created beings*, and thus part of creation! As a result, we will begin to more fully understand that we are image bearers created by God with the specific role to steward creation. Finally, along with our increased awareness that we are part of creation, we will realize that any harm that comes to nature will eventually harm us or our neighbors living in fragile ecosystems that cannot support sustained overuse of resources. This new perspective will gradually help us to cultivate humility in our use and protection of resources, to exercise contentment with consuming only what is needed, and to joyfully enhance our delight in creation, as Revelation 4:11 declares: "Thou art worthy, O Lord, to receive glory and honour and power: for thou hast created all things, and for thy pleasure they are and were created" (KJV).

RESPONSE

Christina Bieber Lake

The most exciting thing about Dr. Page's first chapter is its quiet but forceful argument that if we give up on the liberal arts, we will be giving up on the environment too. And that's because Page is herself no run-of-the-mill, sit-in-the-silo biologist. She is herself a scholar and a teacher, and a perfect example of our deep need for what only the liberal arts can offer, which is an understanding of our interconnectedness with each other, with the natural world, and with Jesus Christ. Page is the most voracious reader of fiction of all the scientists I have met. She is also a gifted photographer. Since she is an artist, she deeply understands that it is art that most powerfully transforms our vision of the world. I am thus delighted to extend a few of her insights here.

I'll begin with Dr. Page's conversation with Bill McKibben. McKibben has had a deep and transformative impact on the fight to defend our environment, and Page explains how he became an advocate. McKibben was deeply moved by the central place given to the natural world by the Inklings. McKibben did not just have cognitive understanding about the issue; he *felt* how C. S. Lewis cherished the forest. This is because McKibben's moral imagination was shaped at an early age by the power of story, particularly the power of story to make us pay attention to what we see (without actually seeing) every day. Without those stories to help him to feel a resonance with the natural world, he would not be the champion of the environment that he is today. Nor would Page.

And nor would I, a non-scientist, have learned to care. After I read this paper, it occurred to me that Page and I are like two versions of the same person. She's the lover of nature and books and art who became a biologist; I'm the lover of nature and books and art who became a literature scholar. This chapter inspired me to remember my earliest adventures in reading, all of which have a thread running through them that is easy to see. Perhaps the earliest book I remember from childhood was Eleanor Thompson's *What Shall I Put in the Hole That I Dig?*, in which the protagonist discovers, among other nonstarters, that when she puts a button in a hole in the ground it does not grow into a button tree.[1] I was later obsessed with Jean Craighead George's classic *My Side of the Mountain*, in which the young protagonist decides to live alone in nature.[2] He learns how to live off of the land—including how to make acorn pancakes. The wildness of this book completely captured my imagination. So when I read in Page's paper that *acorn* was one of the words that the Oxford Junior Dictionary felt it needed to remove to make space for tech-heavy words, I felt the loss of that change more than I might have otherwise. *My Side of the Mountain* was pivotal in my own intellectual formation, for it was a small step from

---

[1]Eleanor Thompson, *What Shall I Put in the Hole That I Dig?* (Atlanta: Whitman, 1963).

[2]Jean Craighead George, *My Side of the Mountain* (New York: Dutton, 1988).

that book to the work of Henry David Thoreau. It was reading and loving *Walden* that turned me into an English major with a focus on American literature.[3] And the most significant connection here is that for Thoreau, to be a poet is to be a naturalist, and to be a naturalist is to be a poet. They are one and the same.

I give this personal information only as background for drawing out and highlighting Page's most important point: that in order to be moved to care for the forest, we must learn how to see it differently. That is work best done by the storyteller and the poet. We must have our imaginations stirred in order to learn to see the forest as Lewis and Tolkien did: as fully alive. We must learn to love the forest in order to see it as endangered by the forces of careless or rapacious industry. And most of all, we must love the forest in order to understand that our flourishing is utterly dependent on the forest's flourishing. We need to see ourselves as tiny hobbits sitting in the branches of the bigger and more powerful Treebeard. Page carefully points out that Christian notions of "subduing the earth" and even of "stewardship" have not generally been persuasive enough on this last point. We must imagine a *different kind of relationship* between ourselves and the forest, which requires art to help us see.

It turns out that, like the Inklings, many American writers have offered us this vision. I'm proposing that we think of ourselves not as stewards but as "interdependent caretakers." This is just my way of phrasing the land ethic that Page mentioned, my way of agreeing with her conclusion that "any harm that comes to nature, will eventually harm us." Many American writers have believed this to be true and have been

---

[3]Henry David Thoreau, *Walden, or Life in the Woods* (1872).

trying to stir our imaginations to cause us to act. Let me give just two of many examples. The first shows the importance of story to fuel our sense of responsibility, the second, of poetry to help us see and feel the beauty of the created world.

When it comes to a novel that stirs us to love the forest, it is difficult to think of one more powerful than Richard Powers's *The Overstory*.[4] In many ways, this award-winning novel is a novel for our age. It is impossible to describe this novel; you really must experience it. It is structured in a way that illustrates our interdependence with the forest: it contains multiple persons' stories interlaced with each other—and with actual trees—like a magnificent root system. The novel fleshes out one of its epigraphs, from the scientist who proposed the Gaia theory of the earth, James Lovelock. Lovelock writes that

> Earth may be alive: not as the ancients saw her—a sentient Goddess with a purpose and foresight—but alive like a tree. A tree that quietly exists, never moving except to sway in the wind, yet endlessly conversing with the sunlight and soil. Using sunlight and water and nutrient minerals to grow and change. But all done so imperceptibly, that to me the old oak tree on the green is the same as it was when I was a child.[5]

You do not need to be a proponent of Gaia to agree with this statement and to understand that the way the earth is alive is bigger than most of us have the capacity to imagine without help. Since twenty-first century urban dwellers lack that capacity, *The Overstory* sinks us into the quiet life of trees

---

[4]Richard Powers, *The Overstory* (New York: W. W. Norton, 2019).
[5]Powers, *Overstory*, epigraph.

and the people—all of us—whose every breath depends on the earth's living forests. It's a remarkable book that, among other things, brings readers to tears over the loss of the American chestnut tree.

There are hundreds of poets I could choose whose work begs us to see and respect the beauty of the natural world and of the forest, but I've chosen Robert Frost, the New England farmer-poet. I taught a class one semester on Frost alone, and after weeks of study a student came up with a brilliant metaphor to describe the relationship that Frost sees between the farmer-poet and the natural world: husbandry. Leaving aside any gender associations and returning to the word's old Norse roots, *husband* designates someone who is master of the house (*hús*) as well as an occupier and tiller of the soil (*bóndi*). Husbandry still carries rather genteel associations of the type of care that emerges when we understand that we need the land more than it needs us—but it does, indeed, need us. The answer to our God-given interdependence is not the fantasy of eliminating all humans and letting nature run wild. When we cultivate the land, when we husband it properly, it also flourishes.

Frost believed that poetic rhythm and meter resonated with the beauty of the created world, including the speech that came naturally to working-class New Englanders. For Frost, to write a good poem is to husband language properly so that it allows the reader see the beauty of the land and the people who dwell within it harmoniously. This becomes especially clear in his early poem, "Mowing," in which the speaker talks about how his scythe whispers as it clears the grasses. This is exactly how a poet revises his or her speech until it sings with the truth it is uncovering. Thus the speaker insists that,

It was no dream of the gift of idle hours,
Or easy gold at the hand of fay or elf:
Anything more than the truth would have seemed too weak
To the earnest love that laid the swale in rows[6]

Note that the speaker insists that it is someone's earnest love that puts the swale—the low marshy place—in rows. This love requires labor; it is not elfin magic that turns straw into gold. Cultivation without love is destruction. Love without cultivation is utopian fantasy.

This same sentiment is expressed in Frost's poem "The Wood-Pile," in which the speaker finds himself lost in the "frozen swamp one gray day," far away from home.

The hard snow held me, save where now and then
One foot went through. The view was all in lines
Straight up and down of tall slim trees
Too much alike to mark or name a place by
So as to say for certain I was here
Or somewhere else: I was just far from home.[7]

The speaker cannot find his place in the landscape, as everything looks the same. He feels homeless until he comes across a solitary cord of maple:

It was a cord of maple, cut and split
And piled—and measured, four by four by eight.
And not another like it could I see.
No runner tracks in this year's snow looped near it.
And it was older sure than this year's cutting,

---

[6]Robert Frost, "Mowing," Poetry Foundation, www.poetryfoundation.org /poems/53001/mowing-56d231eca88cd.
[7]Robert Frost, "The Wood-Pile," Poetry Foundation, www.poetryfoundation .org/poems/44276/the-wood-pile.

Or even last year's or the year's before.
The wood was gray and the bark warping off it
And the pile somewhat sunken. Clematis
Had wound strings round and round it like a bundle.
What held it though on one side was a tree
Still growing, and on one a stake and prop,
These latter about to fall. I thought that only
Someone who lived in turning to fresh tasks
Could so forget his handiwork on which
He spent himself, the labor of his ax,
And leave it there far from a useful fireplace
To warm the frozen swamp as best it could
With the slow smokeless burning of decay.[8]

As the speaker muses about who cut the maple down and why the cord is left out here, "far from a useful fireplace," Frost leads readers to ponder our own husbandry of the natural world. Are we laboring properly? Frost seems to be saying that to be at home on this earth, if only for the short time we have been allotted, we must not continually turn to the allure of "fresh tasks." We must be caretakers who recognize our interdependence with that which we care for. The speaker gently chides the woodcutter for spending his labor on this pile and then forgetting about it, leaving it subject to entropy. Waste is inevitable when we forget the fact of our interdependence with the natural world.

Beside their support for a land ethic, what do Richard Powers and Robert Frost—two otherwise very different writers—share? Both insist that only the power of art will instruct us, via the faculty of the imagination, to see the natural

---

[8]Frost, "The Wood-Pile."

world for everything that it is. This is why Powers's novel is so long. He knows that most contemporary readers, as distracted as we all are, require an immersion of several hours for our moral imaginations to be shaped. Frost also knows that we need to sit with a poem for a long time to see that its beauty is also an argument for the beauty of its subject. Beauty draws us to love, for we can only love something that we have learned to see as beautiful. As Page's paper so brilliantly illustrates, without the power of the writer's imagination drawing us to love what we cannot always see, the natural world will suffer. We will be consigned to paying attention to our carelessness only after it is far, far too late.

A LAMENT FOR CREATION

Responding to the Groaning of God's World

What does it mean to love your neighbor in the twenty-first century? There are more than 7.5 billion people currently living on earth, and every single person needs resources provided by creation. Obviously, we all need water and food just to live, but we also need other things like timber and fiber so that we can have shelter and clothing, and fuel to move us around town or beyond. God in his love has given us not only a beautiful world but one rich with resources that provide all that we need to live life fully and thrive. The problem is that many of us use resources beyond our immediate needs.[1] When we use more than we need, it changes what is available to our neighbor. Have you ever considered that caring for creation is a way of loving one's neighbor?

Consider this example from J. R. R. Tolkien's *The Return of the King* known as the *scouring of the Shire*. In this passage,

---

[1] I gave a TowerTalk that had a similar introduction. See Kristen Page, "The Impact of Creation Care on Public Health," Wheaton College TowerTalks, 2018, www.wheaton.edu/academics/faculty/towertalks/the-impact-of-creation-care-on-public-health/.

the hobbits have returned from their long time away and many harrowing days fighting the evil that was threatening Middle-earth, only to find that this same evil has despoiled the Shire. As Tolkien explains:

> It was one of the saddest hours in their lives. The great chimney rose up before them; and as they drew near the old village across the Water, through rows of new mean houses along each side of the road, they saw the new mill in all its frowning and dirty ugliness: a great brick building straddling the stream, which it fouled with a steaming and stinking overflow. All along the Bywater Road every tree had been felled. As they crossed the bridge and looked up the Hill they gasped. Even Sam's vision in the Mirror had not prepared him for what they saw. The Old Grange on the west side had been knocked down, and its place taken by rows of tarred sheds. All the chestnuts were gone. The banks and hedgerows were broken. Great wagons were standing in disorder in a field beaten bare of grass. Bagshot Row was a yawning sand and gravel quarry. Bag End up beyond could not be seen for a clutter of large huts. "They've cut it down!" cried Sam. "They've cut down the Party Tree!" He pointed to where the tree had stood under which Bilbo had made his Farewell Speech. It was lying lopped and dead in the field. As if this was the last straw Sam burst into tears.[2]

Here we read of an agricultural environment that has been carelessly uprooted in order to make way for a quarry and new mill. Everywhere, trees and plants have been destroyed only to be replaced by bare soil; smoke from the mill's great

---

[2]J. R. R. Tolkien, *The Return of the King: Being the Third Part of The Lord of the Rings* (London: Harper Collins, 2002), 1016.

chimney chokes the air, while the nearby stream has been
fouled by the mill's overflow. The ecosystem that used to
provide sustenance and enjoyment for the residents of Hobbiton has been degraded and will, if left unchecked, become
untenable for most forms of life. The Shire has become a
victim of wanton industrialization without regard for its
natural world. The ecosystem of this corner of Middle-earth
has been broken.

Likewise, in our world, ecosystems are at risk—ecosystems
that were created to provide us with much more than the raw
materials that supply our day-to-day needs. They also provide
numerous services that we take for granted, including climate
regulation, the water cycle, pollination, food production, and
waste elimination. If our technology could re-create the services that ecosystems provide, services created and gener-
ously provided for the good of all by our loving God—it would
cost us trillions of dollars.[3] However, many of us use more
than our share, and the many ways we remove or extract
natural resources from the earth result in large-scale land
transformations, placing significant strains on these valuable
ecosystem services. In fact, overconsumption and degradation of resources results in a $3-5 trillion annual loss to our
economic well-being,[4] and it has negatively impacted nearly
one-half of the global land mass.[5]

Over the years, many authors have tried to get our attention and move us toward a change in these behaviors.
Rachel Carson is a marine biologist known for one of the

---

[3]David Holzmann, "Accounting for Nature's Benefits: The Dollar Value of Ecosystem Services," *Environmental Health Perspectives* 120 (2012): A153-A157.
[4]Holzmann, "Accounting for Nature's Benefits," A153-A157.
[5]Peter Vitousek, Harold Mooney, Jane Lubchenco, and Jerry Melillo, "Human
Domination of Earth's Ecosystems," *Science* 277 (1997): 494-99.

earliest and most influential environmental books, *Silent Spring*. In the introduction to her book, she offers a "Fable for Tomorrow" in which she describes the transformation of a beautiful landscape of "prosperous farms, with fields of grain and hillsides of orchards [with] white clouds of bloom drift[ing] above the green fields" to a place where animals died, illness plagued people, and "everywhere was a shadow of death."[6] *Silent Spring* was a wakeup call to many about the actual costs of "better living through chemistry"—the promise made by DuPont in a 1930s advertising campaign.[7] In contrast to the unfounded promises of this industrial giant, Carson describes how the uninhibited use of chemicals, particularly DDT, has consequences for ecosystems and human health. She argued that in an age of increasing industrialization, "the right to make a dollar at whatever cost [is] seldom challenged."[8] This assumption can in turn lead to patterns of chemical use without testing, and ultimately to an obfuscation of facts about the implications of these chemicals on the environment and health. Carson not only had the attention of the American public, but she also testified before Congress in 1963 and warned legislators about the dangers of the unregulated use of chemicals on "water, soil, air and vegetation . . . even . . . bodies of animals and . . . men."[9] In the same year, the Clean Air Act became law, and the Water Quality Act followed in 1965, becoming law in 1977.

[6]Rachel Carson, *Silent Spring* (New York: Houghton Mifflin Harcourt, 1962), 1-4.
[7]Livia Gershon, "What We Mean by 'Better Living,'" JSTOR *Daily* (2019), https://daily.jstor.org/what-we-mean-by-better-living/.
[8]Carson, *Silent Spring*, 14.
[9]Rachel Carson, "Environmental Hazards Control of Pesticides and Other Chemical Poisons," statement of Rachel Carson before the Subcommittee on Reorganization and International Organizations of the Committee on Government Operations, June 4, 1963, https://rachelcarsoncouncil.org/about-rcc/about-rachel-carson/rachel-carsons-statement-before-congress-1963/.

Even with expanding environmental legislation and regulations, it is still true that "the burden of proof is on proving a chemical is dangerous rather than on the side of those who introduce the chemical to prove that it is safe."[10] This has resulted in great harm to many communities, ranging from adverse health effects of PCB-contaminated breast milk of the Akwesasne peoples living along the industrialized St. Lawrence River to cancers and other diseases among those living in Parkersburg, West Virginia, linked to the contamination of the Ohio River with perfluorooctanoic acid (C8) from the production of Teflon—a chemical that incidentally can also be found in most homes in the United States.[11] Nearly sixty years after the Clean Air Act, many environmental regulations are not enforced,[12] and in 2020, more than one hundred of them were overturned for the sake of an improved economy.[13] However, the following questions and more remain unanswered. What does it mean to reverse these regulations? At what cost to the health of the land and our bodies do we ignore the damage done to creation by unregulated actions? What are the implications for our future? Are we gaining short-term economic benefit while squandering the long-term economic

[10]Eric Chivian as quoted in Gershon, "What We Mean by 'Better Living.'"

[11]Winona LaDuke, *All Our Relations: Native Struggles for Land and Life* (Cambridge: South End, 1999); Taylor Sisk, "A Lasting Legacy: Dupont, C8 Contamination and the Community of Parkersburg Left to Grapple with the Consequences," *Environmental Health News*, January 7, 2020, www.ehn.org /dupont-c8-parkersburg-2644262065.html.

[12]Oliver Milman and Emily Holden, "Trump Administration Allows Companies to Break Pollution Laws During Coronavirus Pandemic," *The Guardian*, March 27, 2020, www.theguardian.com/environment/2020/mar/27/trump -pollution-laws-epa-allows-companies-pollute-without-penalty-during -coronavirus.

[13]Nadja Popovic, Livia Albeck-Ripka, and Kendra Pierre-Louis, "The Trump Administration Is Reversing 100 Environmental Rules: Here's the Full List," *New York Times*, July 15, 2020, www.nytimes.com/interactive/2020/climate /trump-environment-rollbacks-list.html.

services that God built into our ecosystems? Are we, perhaps, living in Rachel Carson's "Fable for Tomorrow?" And if so, how do we best address these questions?

Many who love the fiction of Lewis and Tolkien believe that storytelling is a critical means of transforming attitudes, including ecological choices.[14] According to Langellier and Peterson, "People make sense of their experiences, claim identities, interact with each other, and participate in cultural conversations through storytelling."[15] Stories are important because the listener is in a relationship with the storyteller and thereby experiences events in authentic ways. Such experiences can result in transformation of attitudes, as if the events of the story actually happened in the life of the reader. C. S. Lewis explains that we read "to be more than ourselves . . . to see with other eyes, to imagine with other imaginations, to feel with other hearts, as well as with our own."[16]

Describing a need for a change of heart within the context of familiar places is an effective means of changing the reader's heart. Lewis and Tolkien both created worlds with places that readers know well and care about. For those of you who love Narnia, do you remember how you felt when winter began to end, and you heard "chattering and chirruping in every direction, and then a moment of full song, and within five minutes the whole wood (was) ringing with birds'

---

[14]Steven Bouma-Prediger, *Earthkeeping and Character: Exploring a Christian Ecological Virtue Ethic* (Grand Rapids, MI: Baker Academic, 2020); Matthew Dickerson and David O'Hara, *Narnia and the Fields of Arbol: The Environmental Vision of C. S. Lewis* (Lexington: University Press of Kentucky, 2008); Matthew Dickerson and Jonathan Evans, *Ents, Elves, and Eriador: The Environmental Vision of J. R. R. Tolkien* (Lexington: University Press of Kentucky, 2006).

[15]Kristin Langellier and Eric Peterson, *Storytelling in Daily Life: Performing Narrative* (Philadelphia: Temple University Press, 2004).

[16]C. S. Lewis, *An Experiment in Criticism* (London: Cambridge University Press, 1961), 137-38.

music"?[17] Did you notice that the thaw in Edmund's heart co-incided with the breaking up of winter? I wonder how many readers see themselves in Edmund and realize their need for spring and a new beginning as well.

Tolkien describes the way that stories are able to transform attitudes as "recovery." In his essay "On Fairy-Stories," he explains that "recovery . . . is a re-gaining—regaining of a clear view."[18] If you have ever read Tolkien's *Lord of the Rings*, perhaps you "regain[ed] a clear view," or experienced the scouring of the Shire in a way that caused you to recognize something about your own attitudes toward consumption. How did you feel when Frodo, Sam, Merry, and Pippin returned to the Shire and were met with this devastating situation? The place that they love has been transformed from a comfortable and hospitable land that provided for the needs and wants of the hobbits to an unrecognizable place that is characterized by burning and smoke.[19] "Gloomy" and "un-Shirelike" houses were built where trees and hedgerows had been.[20] The group of friends were not greeted with a beer and a pipe as they expected but rather discovered rules that benefited foreigners.[21] They were confronted by a radical transformation of the land that they loved—"This was Frodo and Sam's own country, and they found out now that they cared about it more than any other place in the world."[22] In fact, they cared about it so much that they acted! "Wake all our people! They hate all this, you

---

[17]C. S. Lewis, *The Lion, the Witch, and the Wardrobe* (New York: Collier, 1950), 117.

[18]Verlyn Flieger, and Douglas A. Anderson, eds., *Tolkien on Fairy-Stories* (London: HarperCollins, 2008).

[19]Tolkien, *The Return of the King*, 1000.

[20]Tolkien, *The Return of the King*, 998.

[21]Tolkien, *The Return of the King*, 1000.

[22]Tolkien, *The Return of the King*, 1004.

can see: all of them except perhaps one or two rascals, and a few fools that want to be important, but don't at all understand what is really going on. But Shire-folk have been so comfortable so long they don't know what to do."[23]

The changes leading up to the situation that Frodo and Sam found on their return to the Shire occurred slowly. Many of the attitudes that ultimately resulted in the "Scouring of the Shire" existed before the Fellowship even began their journey with the ring. Even Frodo and friends, prior to their journey, shared in the complacency of comfort; however, the distance (both physical and emotional) during their journey allowed them to see, on their return, the harm of such attitudes. Literature does this for us. As we read, we can recognize our own stories and possibly even begin to envision a response to the injustices we discover.

While stories may awaken us to the environmental distress of our world and even to our own sense of loss as nature is harmed, in order to find solutions, we must better understand the problems we are facing. As Jesus instructed, we must learn to "love the Lord your God with all your heart and with all your soul and with all your mind" (Mt 22:37). This call to use our mind is where science can best help us as we thoughtfully examine the documented studies of environmental experts. Like the hobbits living in the Shire, many of us have been living in "comfort for so long that [we] do not know what to do."[24] However, our comfort comes at the expense of many of our global neighbors, who suffer as ecosystems are degraded. For example, if everyone on earth used resources at the same rate as the average American, we would need

[23]Tolkien, *The Return of the King*, 1007.
[24]Tolkien, *The Return of the King*, 1007.

more than five earths.[25] In fact, we use resources so quickly, we consistently reach the annual limit of what the earth can naturally regenerate months before the end of the year. Earth Overshoot Day is the day each year when we reach our ecological budget, begin to "overspend," and use resources that are not easily regenerated (if they can be regenerated at all).[26] In 1970, this date was December 29, but each subsequent decade brought earlier and earlier Earth Overshoot Days. By the end of the 1980s, resource use exceeded regeneration by the second week of October, and we reached Earth Overshoot Day by early September in the early 2000s.[27] In 2019, Earth Overshoot Day was July 29,[28] but because of the global coronavirus pandemic in 2020, it was three weeks later than previous years. We can thank the need for most of the world to stay home to prevent the spread of the virus for an 8.4 percent reduction in the use of forest products and a 14.5 percent reduction in our carbon footprint.[29] You might be quick to celebrate this as a silver lining, but if we hurry to celebration, we might miss the warning. The remarkable change in trajectory that has resulted from most of the world being forced to stay home should illustrate the impact of our "normal" levels of consumption. In fact, the trend of earlier and earlier Earth Overshoot Days likely contributed to the emergence of

[25]C. McDonald, "How Many Earths Do We Need?" BBC News Magazine (2015), www.bbc.com/news/magazine-33133712.

[26]Global Footprint Network, "How the Date of Earth Overshoot Day 2020 Was Calculated," Earth Overshoot Day, June 5, 2020, www.overshootday.org /2020-calculation/.

[27]Global Footprint Network, "Past Earth Overshoot Days," Earth Overshoot Day, 2021, www.overshootday.org/newsroom/past-earth-overshoot-days/.

[28]Tessa Koumoundouros, "We Just Used Up All of Earth's Resources for the Year, and It's Only July," Science Alert, July 29, 2019, www.sciencealert.com /we-just-used-up-all-of-earth-s-resources-for-the-year-and-it-s-only -july.

[29]Global Footprint Network, "Earth Overshoot Day."

this global pandemic. Our persistent overuse of resources, and the impact of our consumption on the environment, is directly linked to the emergence of diseases like the novel coronavirus (SARS-CoV-2).[30] For example, in a span of less than fifty years at the end of the twentieth century, more than forty new diseases emerged across the globe.[31]

Those of us with the means to consume resources beyond our immediate needs are often the most protected from the impacts of our resource use on the environment. We rarely see environments that lead to sickness because we are sheltered from environmental harm by zoning or environmental laws that protect wealthy neighborhoods within wealthy countries.[32] However, when we stand with Frodo, Sam, and Gollum at the gates of Mordor, we can experience what many people living in poverty experience daily—environments ravaged by the dumping of toxins and the uninhibited extraction of resources that lay waste to the land:

> Here neither spring nor summer would ever come again. Here nothing lived, not even the leprous growths that feed on rottenness. The gasping pools were choked with ash and crawling muds, sickly white and grey, as if the mountains had vomited the filth of their entrails upon the lands about. High mounds of crushed and powdered rock, great cones of earth fire-blasted and poison-stained,

---

[30]Jonathan Foley et al., "Global Consequences of Land Use," Science 309 (2005): 570-74.

[31]"A Safer Future: Global Public Health Security in the Twenty-first Century," World Health Organization, 2007, https://apps.who.int/iris/bitstream /handle/10665/43713/9789241563444_eng.pdf?sequence=1&isAllowed=y.

[32]For example, see descriptions of communities associated with brownfields in Maryland or those associated with coal-fired power plants in Chicago in Noah Toly, The Gardeners' Dirty Hands: Environmental Politics and Christian Ethics (New York: Oxford University Press, 2019).

stood like an obscene graveyard in endless rows, slowly revealed in the reluctant light.

They had come to the desolation that lay before Mordor: the lasting monument to the dark labour of its slaves that should endure when all their purposes were made void; a land defiled, diseased beyond all healing— unless the Great Sea should enter in and wash it with oblivion. "I feel sick," said Sam. Frodo did not speak.[33]

This passage reminds me of a place I worked in the mid-1990s when I was an intern with an environmental consulting firm. I was tasked with conducting environmental sampling of an industrial site in an urban center to determine the types of environmental contaminants and whether the contamination was at levels that required remediation. At the time, the site was occupied by a food production company, but it had formerly been the location of a chemical company and was being considered for purchase by a third company. This location is considered a *brownfield*—a former industrial site that is potentially contaminated with chemicals—and it was the most disgusting place I have ever been. The smell was nauseating, and every few steps there were piles of what looked like chalky white and gooey green chemical residue: "The gasping pools were choked with ash and crawling muds, sickly white and grey."[34] I do not know the ultimate results of my sampling, but I do remember that every day I spent at that location, I felt sick like Sam. Like Sam, I was able to leave and return to my comfortable home, yet many people do not have the option to leave and must live near urban brownfields like the one I described.

---

[33]J. R. R. Tolkien, *The Two Towers: Being the Second Part of The Lord of the Rings* (London: HarperCollins, 1954), 631-32.
[34]Tolkien, *Two Towers*, 631.

In fact, the EPA estimates that there are as many as one million brownfield sites in the United States, and 85-90 percent of these have not yet been evaluated for remediation.[35] Because most brownfields are located in urban areas, they disproportionately impact minorities and low-income families, whose children are the most vulnerable.[36] Tragically, as many as 25 percent of children living in the United States live within two kilometers of a brownfield, and because the most common contaminants in brownfields are genotoxic (damaging to genetic material), these children have increased cancer risk.[37] It is estimated that the risk of toxic exposure increases ten times for children under the age of two.[38] This is especially serious when you consider that industry in these areas is less likely to be inspected for environmental infractions, the EPA is less likely to enforce compliance,[39] and children living in neighborhoods close to these brownfields do not have equal access to healthcare.[40]

---

[35]US Environmental Protection Agency, "Cleaning Up the Nation's Waste Sites," National Service Center for Environmental Publications, 2004, https://bit.ly/3KxbZzE.

[36]Kristin Shrader-Frechette and Andrew Biondo, "Protecting Children from Toxic Waste: Data-Usability Evaluation Can Deter Flawed Cleanup," *International Journal of Environmental Research and Public Health* 17 (2020): 424-59.

[37]Shrader-Frechette and Biondo, "Protecting Children from Toxic Waste." Exposures to genotoxic chemicals are more dangerous for children as they are growing and experiencing rapid rates of cell division.

[38]Shrader-Frechette and Biondo, "Protecting Children from Toxic Waste." During rapid growth there is a high rate of cell division—mutagens can result in errors in cell division that result in cancers.

[39]Wayne Gray, Ronald Shadbegian, and Ann Wolverton, "Environmental Justice: Do Poor and Minority Populations Face More Hazards?" National Center for Environmental Economics, Working Paper #10-10, US Environmental Protection Agency, 2010, www.epa.gov/sites/default/files/2014-12/documents/environmental_justice_do_poor_and_minority_populations_face_more_hazards.pdf.

[40]Dolores Acevedo-Garcia et al., "Racial and Ethnic Inequities in Children's Neighborhoods: Evidence from the New Child Opportunity Index 2.0," *Health Affairs* 39, no. 10 (2020): 1693-701.

It might not be a surprise that pollution with toxic chemicals makes people sick; however, most of us live comfortably away from these issues, so we may not often think about the dangers of pollution. We live in tidy homes with sanitation systems, and landfills keep us separate from our waste and other conditions we associate with diseases. In fact, in the developed world we live in a way that keeps us separate from nature altogether. We may visit for recreation, but we do not believe that our use of resources and the way we interact with our environment could possibly make us (or our neighbors) sick. Our highly developed and constructed landscapes perpetuate the false idea that we can keep ourselves distanced from nature and the primary cause of infectious disease: germs. With discoveries by scientists like Louis Pasteur and Robert Koch that led to the development of germ theory,[41] the identified pathogens could be controlled through vaccines or medications. No longer did people associate disease with uncontrollable phenomena like stars and weather events; rather, scientific understanding gives us more control over the outcomes of disease. Controlling the environment to prevent disease ultimately resulted in a perceived dichotomy, with humans separate from and in control of the rest of creation. More than 150 years after the publication of germ theory, we find it hard to accept that germs, the environment, and our actions are linked at all. Aldo Leopold, an ecologist writing in the 1930s, sees this perceived separation of humans from nature as a causative factor of "land sickness,"[42] and his writings urge us to consider that we too are part of nature. He

---

[41]Gregg Mitman, "In Search of Health: Landscape and Disease in American Environmental History," *Environmental History* 10, no. 2 (2005): 184-210.

[42]Aldo Leopold, *The River of the Mother of God and Other Essays*, ed. Susan Flader and J. Baird Callicott (Madison: University of Wisconsin Press, 1991).

insists that a change in attitudes is imperative to the health of the land, and by extension, the health of the inhabitants of the land. Sadly, we did not heed the warnings of Leopold, and the pursuit of resources drastically changed our landscapes to the point that by the end of the twentieth century, many ecologists began to produce scholarship with overwhelming evidence that human-driven changes in landscape alter ecological factors that are involved in the emergence of diseases.[43] In fact, most diseases have emerged after human actions significantly changed ecosystems, and human behavior brought us closer to animal reservoirs of pathogens.[44] As we entered the twenty-first century, a comprehensive study reported 1,415 species of pathogenic organisms that range from bacteria to viruses and fungi to worms. 175 pathogens are considered emerging, and of these, more than 75 percent are zoonotic (transmitted from animals to humans).[45] SARS-CoV-2 is not included in these totals because it is too new, but it too is a zoonotic virus that spilled over into human populations as a result of the ways we extract and consume resources. In fact, scientists and world leaders involved in the G7 Climate and Environment response have recently recognized the importance of intact landscapes, coined *landscape immunity*, for protection from emerging diseases including SARS-CoV-2.[46]

---

[43]Kris Murray and Peter Daszak, "Human Ecology in Pathogenic Landscapes: Two Hypotheses on How Land Use Change Drives Viral Emergence," *Current Opinion in Virology* 3 (2013): 79-83.

[44]Raina Plowright et al., "Causal Inference in Disease Ecology: Investigating Ecological Drivers of Disease Emergence," *Frontiers in Ecology and the Environment* 6, no. 8 (2008): 420-29.

[45]Louise Taylor, Sophia Latham, Mark Woolhouse, "Risk Factors for Human Disease Emergence," *The Philosophical Transactions of the Royal Society of London* 356 (2001): 983-89.

[46]Jamie Reaser, Brooklin Hunt, Manuel Ruiz-Aravena, Gary Tabor, Jonathan Patz, Daniel Becker, Harvey Locke, Peter Hudson, and Raina Plowright, "Fostering Landscape Immunity to Protect Human Health: A Science-Based

Most zoonotic diseases that spill over into human populations disproportionately affect the poor, who are nutritionally stressed and typically live in ecologically vulnerable parts of the world—places where ecosystems are slow to recover from resource extraction due to fragile soils and pathogens benefit from warm and humid climates. For example, Ebola has emerged more than twenty times across the African continent since 1976 as the demand for minerals, timber and agricultural land has transformed the tropical landscape.[47] This landscape transformation has resulted in a decreasing natural biodiversity and has subsequently placed humans in contact with the remaining wildlife reservoirs of the disease. Most of us have heard about the 2014 outbreak of Ebola in Guinea, Liberia, and Sierra Leone, resulting in 28,000 cases that claimed more than 11,000 lives, but there have also been twenty-seven outbreaks resulting in 34,000 cases and the loss of 15,000 lives over the past forty-four years.[48] There is strong evidence that Ebola emerges within two years of a deforestation event, providing a compelling argument to protect forests and biodiversity.[49] Even amid our current pandemic of SARS-CoV-2, which we strongly suspect to have emerged from wildlife[50], many of us do not understand how our

Rationale for Shifting Conservation Policy Paradigms," *Conservation Letters* (2022), e12869.

[47]"Years of Ebola Virus Disease Outbreaks: 40 Years of Ebola Virus Disease Around the World," CDC, 2020, www.cdc.gov/vhf/ebola/history/chronology .html.

[48]CDC, "Years of Ebola Virus Disease Outbreaks."

[49]Jesús Oliviero et al., "Recent Loss of Closed Forests Is Associated with Ebola Virus Disease Outbreaks," *Scientific Reports* 7 (2017): 14291, https://doi .org/10.1038/s41598-017-14727-9.

[50]Najmul Haider, Peregrine Rothman-Ostrow, Abdinasir Yusuf Osman, Liã Bárbara Arruda, Laura Macfarlane-Berry, Linzy Elton, Margaret Thomason, Dorothy Yeboah-Manu, Rashid Ansumana, Nathan Kapata, Leonard Mboera, Jonathan Rushton, Timothy McHugh, David Heymann, Alimuddin Zumla, and

Plate 1. Deodar cedar (*Cedrus deodara*) are common in the forests above Landour, India, in the state of Uttarakhand. This high-altitude (>7000ft) forest is in the foothills of the Himalayas. Photo by Kristen Page.

Plate 2. Ancient Himalayan Oak (*Quercus leucotrocophora*) growing in the forests above Landour, India, in the state of Uttarakhand. Photo by Kristen Page.

Plate 3. Windmills at sunset at South Padre Island, Texas, remind us to reconsider the ways we use natural resources to lessen the negative impacts on creation and our neighbors. Photo by Kristen Page.

Plate 4. Snow Geese (*Chen caerulescens*) take flight on their wintering grounds at Hagerman National Wildlife Refuge in North Texas. Photo by Kristen Page.

Plate 5. Young lion (*Panthera leo*) in Maasai Mara National Reserve, Kenya. Photo by Kristen Page.

Plate 6. Under the Milky Way in rural Illinois. Photo by Kristen Page.

Plate 7. Elephant (*Loxodonta africana*) in Tarangire National Park, Tanzania. Photo by Kristen Page.

Plate 8. Mountain Gorillas (*Gorilla gorilla*) in Bwindi Impenetrable National Park, Uganda. There are just over 1,000 mountain gorillas remaining. Photo by Kristen Page.

individual behavior could possibly be related to the outbreak of diseases on distant continents. It would be easy to blame those living in the place where the outbreak emerged, in this case Wuhan, China. We could make the judgment that the landscape changes that enabled the spread of SARS-CoV-2 reflect only local decisions regarding natural resources; however, global actions, including our own, are driving the current large-scale landscape changes that accompany disease emergence. In other words, our consumption of everything from chocolate to coffee causes large-scale changes in the world's landscape. For example, the incredibly diverse rainforests of West Africa, where Ebola emerged in 2014, have steadily been cut due to the expansion of cocoa plantations;[51] as a result, forests have been reduced by nearly 30 percent since the 1970s, when Ebola first emerged on the African continent.[52] Thus, our insatiable appetite for chocolate and coffee plays a role in the transformation of landscapes,[53] which not only facilitates the emergence of diseases like Ebola but also accelerates global climate change.[54] Many diseases, like malaria, dengue, and Zika, are expanding in range and becoming a greater threat to more people as a result of increasing global

---

Richard Kock, "Covid-19-Zoonosis or Emerging Infectious Disease?" *Frontiers in Public Health* 8 (2020), 596944.

[51]François Ruf, Götz Schroth, and Kone Doffangui, "Climate Change, Cocoa Migrations and Deforestation in West Africa: What Does the Past Tell Us About the Future?" *Sustainability Science* 10 (2015): 101-11.

[52]"West Africa: Land Use and Land Cover Dynamics—The Deforestation of the Upper Guinean Forest in Landscapes of West Africa," USAID and USGS, 2016, https://eros.usgs.gov/westafrica/land-cover/deforestation-upper -guinean-forest.

[53]Laura García-Herrero, Fabio De Menna, and Matteo Vittuari, "Sustainability Concerns and Practices in the Chocolate Life Cycle: Integrating Consumers' Perceptions and Experts' Knowledge," *Sustainable Production and Consumption*, 20 (2019), 117-27.

[54]Foley, et al., "Global Consequences of Land Use," 570-74.

temperatures and the subsequent expansion of mosquito ranges. Likewise waterborne diseases, including cholera, are expanding their reach as oceans warm and sea levels rise. Cholera, caused by the bacterium *Vibrio cholerae*, is one of the deadliest waterborne diseases with a mortality rate of up to 60 percent if untreated. An outbreak of Cholera in Yemen that began in 2016 is one of the largest and fastest-spreading outbreaks in modern history; it has resulted in more than a million cases and 1,500 deaths over the past four years.[55]

Can the emergence of thousands of diseases be a much-needed wakeup call to our world? Will this indication that we are at a tipping point underscore the reality of our own self-interest and need to respond before more pandemics occur?[56] Like the four hobbits returning to the Shire, we have experienced the devastation of SARS-CoV-2. But will we also be courageous enough to follow the example of the hobbits and act in order to make necessary changes to prevent the source of our own devastation? Fifty years after Lynne White published his assertion that Judeo-Christian values have led to an environmental crisis,[57] our worldwide crisis is worsening, the wealthy continue to seek comfort through consumption of more and more resources, and Christians are the most vocal climate change skeptics and deniers.[58] I lament this response,

---

[55]"Epidemic and Pandemic Prone Diseases: Outbreak Update—Cholera in Yemen 12 January 2020," World Health Organization, 2020, www.emro.who.int/pandemic-epidemic-diseases/cholera/outbreak-update-cholera-in-yemen-12-january-2020.html.

[56]Tipping points are considered human impacts on ecological systems that move them beyond their "normal operation." For example, see Timothy Lenton et al., "Tipping Elements in the Earth's Climate System," PNAS 105 (2008): 1786-93.

[57]Lynn White, "The Historical Roots of Our Ecological Crisis," *Science* 155 (1967): 1203-7.

[58]Robin Globus Veldman, *The Gospel of Climate Skepticism* (Oakland: University of California Press, 2019), 2.

or lack thereof. Those of us who make the biggest impact on the environment through consumption are the least likely to be impacted by degrading ecological systems. Cynthia Moe-Lobeda writes that "climate change may be the most far-reaching manifestations of white privilege and class privilege to face humankind. Caused overwhelmingly by high consuming people, climate change is wreaking death and destruction foremost on impoverished people, who are disproportionately people of color."[59] Significantly, these results are not immediately noticed by those of us who are causing the change, rather the "delayed destruction that is dispersed across time and space" is what Rob Nixon has coined "slow violence."[60]

Changes in climate have radically shifted in a short amount of time, and for the people living in "climate privileged" regions, it may be difficult to link actions with outcomes. A slow shift in priorities over generations has caused complicit participation in a culture driven by consumerism. However, though these actions may be unintended, they nonetheless harm our planet as they deplete the ability of ecosystems to replenish and renew. Further, any actions to slow environmental damage have been hindered by our reliance on and comfort in convenient consumerism. For example, our dependence on the internal combustion engine has also increased our reliance on fossil fuels and accelerated changes in climate in a relatively short period of time. The internal combustion engine was invented in the last half of the nineteenth century, and by the early twentieth century, cars were widely

---

[59]Cynthia Moe-Lobeda, "Climate Change as Climate Debt: Forging a Just Future," *Journal of the Society of Christian Ethics* 36, no. 1 (2016): 27-49.

[60]Rob Nixon, *Slow Violence and the Environmentalism of the Poor* (Cambridge: Harvard University Press, 2011).

accessible. Today, 18 percent of the world's population has an automobile,[61] and most of these are concentrated in the United States and China—countries that are also contributing the most to carbon emissions, and thus to climate change.[62] At the time these engines were invented, and for many years after, it was assumed that the CO_2 and water vapor produced as a byproduct of combustion was not an environmental problem. After all, CO_2 and water are harmless to people, and it was believed that any excess CO_2 would just be absorbed by the ocean. Those producing internal combustion engines may have made the assumption, but even before the engine was invented, Eunice Newton Foote conducted some elegant experiments that demonstrated that increasing CO_2 in an experimental atmosphere under glass increases the temperature of the atmosphere.[63] In 1856, when she conducted these experiments, the results were mostly ignored.[64] It took a hundred years before scientists first measured increased CO_2 accumulation in the Earth's atmosphere.[65] While CO_2 is harmless to people, the molecule traps heat in the atmosphere

[61]Andrew Chesterton, "How Many Cars Are There in the World?" Carsguide. com, 2018, www.carsguide.com.au/car-advice/how-many-cars-are-there -in-the-world-70629.

[62]Johannes Friedrich, Mengpin Ge, and Andrew Pickens, "This Interactive Chart Explains World's Top 10 Emitters, And How They've Changed," *World Resources Institute*, 2017, www.wri.org/blog/2017/04/interactive-chart -explains-worlds-top-10-emitters-and-how-theyve-changed; "How Many Registered Motor Vehicles Are There in the U.S.?" Statista, 2020, www.statista .com/statistics/183505/number-of-vehicles-in-the-united-states-since -1990/.

[63]Ayana Johnson and Katharine Wilkinson, *All We Can Save* (New York: Random House, 2020), xvii.

[64]Eunice Foote, "Circumstances Affecting the Heat of the Sun's Rays," *American Journal of Art and Science* 22, no. 66, 2nd series (1856): 382-83.

[65]Charles Keeling, "The Concentration and Isotopic Abundances of Atmospheric Carbon Dioxide in Rural Areas," *Geochimica et Cosmochimica Acta* 13 (1957): 322-34; Daniel Harris, "Charles David Keeling and the Story of Atmospheric CO_2 Measurements," *Analytical Chemistry* 82 (2010): 7865-70.

in a process known as the greenhouse effect. Other greenhouse gasses are important, but CO_2 is one of the most discussed because it persists for the longest duration, is the most abundant, and is among the easiest to measure. In the 1960s scientists began continuous measurements of rising CO_2 levels,[66] began discussing measurable changes to climate,[67] and linked changes in climate to changes in carbon dioxide.[68] In 1988 the first scientist (James Hansen of NASA) went before Congress to warn of increasing temperatures and changing climates,[69] and in 1995, the Intergovernmental Panel on Climate Change (UN) published their first report concluding the planet is warming, and humans are causing it through activities like driving automobiles.[70] Today, the hottest regions (mean annual temperature >29°C) of the earth are currently restricted to less than 1 percent of the earth's surface, but it is predicted that over the next fifty years, this area will increase nearly 20 percent and displace billions of people.[71] Who will this impact the most? Sadly, those who live in poorer regions of the world and are in ecosystems that are fragile and easily harmed. In contrast, the "climate privileged" live in ecologically resilient locations, primarily northern countries in

---

[66]Charles Keeling, "The Concentration and Isotopic Abundances of Carbon Dioxide in the Atmosphere," *Tellus* 12, no. 2 (1960): 200-203.

[67]I. Schell, "Recent Evidence About the Nature of Climate Change and Its Implications," *Annals of the New York Academy of Sciences* 95, no. 1 (1961): 251-70.

[68]F. Möller, "On the Influence of Changes in the CO_2 Concentration in Air on the Radiation Balance of the Earth's Surface and on the Climate," *Journal of Geophysical Research* 68, no. 13 (1963): 3877-86.

[69]Guy Darst, "AP Was There: The Age of Climate Change Begins," *AP News*, 2018, https://apnews.com/59db44d726fa4a608987e674e6f13dd8.

[70]IPCC, "IPCC Second Assessment: Climate Change 1995: A Report on the Intergovernmental Panel on Climate Change," United Nations, 1995.

[71]Abrahm Lustgarten, "The Great Climate Migration," *The New York Times*, September 2, 2020, www.nytimes.com/interactive/2020/07/23/magazine/climate-migration.html.

temperate ecosystems, which also corresponds to high economic development.[72] However, the levels of consumption achieved by the "climate privileged" causes devastating changes to climate systems that ultimately challenge the health of our neighbors in the developing world. It might be hard for us to understand a seemingly small increase in temperature of 1.5°C to be a crisis. We might not even imagine that such a seemingly small temperature change could result in devastating changes to rainfall patterns, surface temperatures, and intensity of storms in other parts of the world. However, this is the reality, and if we do nothing about climate change, in the next fifty years there will be more than 143 million climate refugees throughout sub-Saharan Africa, South Asia, and Latin America. If we act now, we can reduce the number of displaced neighbors by 80 percent.[73]

What we consider *normal* use of resources is increasing each year. Wendell Berry describes our "abundant consumption" as an "indulgence in depend[ence] on, and wastefulness of our economy," which is "causing [a] crisis."[74] Perhaps you have never thought of the dangers to yourself as well as to your neighbors of "abundant consumption." Once again, we can turn to a story to help us apprehend this better. In the second volume of his science fiction trilogy, *Perelandra*, C. S. Lewis describes the surprise of his protagonist Ransom when he first contemplates restraint in enjoying natural resources. As Lewis writes:

---

[72]Lustgarten, "Great Climate Migration."

[73]Kanta Kumari Rigaud et al., "Groundswell: Preparing for Internal Climate Migration," World Bank, 2018, https://openknowledge.worldbank.org /handle/10986/29461.

[74]Wendell Berry, *Think Little* (Berkeley, CA: Counterpoint, 2019), 8, 58.

Now [Ransom] had come to a part of the wood where great globes of yellow fruit hung from the trees—clustered as toy-balloons are clustered on the back of the balloon-man and about the same size. He picked one of them and turned it over and over. The rind was smooth and firm and seemed impossible to tear open. Then by accident one of his fingers punctured it and went through into coldness. After a moment's hesitation he put the little aperture to his lips. He had meant to extract the smallest, experimental sip, but the first taste put his caution all to flight. It was, of course, a taste, just as his thirst and hunger had been thirst and hunger. But then it was so different from every other taste that it seemed mere pedantry to call it a taste at all. It was like the discovery of a totally new *genus* of pleasures, something unheard of among men, out of all reckoning, beyond all covenant. For one draught of this on earth wars would be fought and nations betrayed. It could not be classified. He could never tell us, when he came back to the world of men, whether it was sharp or sweet, savory or voluptuous, creamy or piercing. "Not like that" was all he could ever say to such inquiries. As he let the empty gourd fall from his hand and was about to pluck a second one, it came into his head that he was now neither hungry nor thirsty. And yet to repeat a pleasure so intense and almost so spiritual seemed an obvious thing to do. His reason, or what we commonly take to be reason in our own world, was all in favor of tasting this miracle again; the childlike innocence of fruit, the labors he had undergone, the uncertainty of the future, all seemed to commend the action. Yet something seemed opposed to this "reason." It is difficult to suppose that this opposition came from desire, for what desire would turn from so much deliciousness? But for whatever cause, it appeared to him better not to taste again. Perhaps the experience had been so complete that repetition would be a vulgarity—like asking to hear the same symphony twice in a day. As he stood pondering over this and wondering how often in his life on earth he had reiterated pleasures not through desire, but in the teeth of desire and in obedience to a spurious rationalism.[75]

Does hearing the above passage cause you to consider that God intends that we use restraint in enjoying the natural resources that he has provided for all who live on this earth? In raising this point, Lewis is asking us to think about more than excess indulgence; he is offering us a glimpse into right and proper usage as well as respect for others who are our neighbors in this world. With this in mind, how do we reconcile the disproportionate impacts of the ways we use resources and the environmental degradation that results? Once we recognize the injustice that has resulted from these actions, we are invited to lament. According to Soong-Chan Rah, this is a way for us to "recognize the struggles of life and cry out for justice against existing injustices."[76] According to Rah, "Lament calls for an authentic encounter with the truth and challenges privilege, because privilege would hide the truth that creates discomfort."[77] Lament allows us to confront the truths we have ignored for the sake of comfort and gives

---

[75]C. S. Lewis, *Perelandra* (New York: Scribner, 1996), 42.
[76]Soong-Chan Rah, *Prophetic Lament* (Downers Grove, IL: InterVarsity Press, 2015), 23.
[77]Rah, *Prophetic Lament*, 58.

those of us living in comfort a means to petition God on behalf of our neighbors in need. Rah cautions that lament is not merely a way to complain, but it becomes a true "expression of sadness" and a response to and acknowledgment of the suffering of our neighbor before God.[78]

Lament is the bringer of change, but how can that happen when many evangelicals do not link the suffering of neighbors with their personal consumption patterns and attitudes toward creation? Evangelicals "are the most skeptical major religious group in the country when it comes to climate change,"[79] and as a result the least likely to realize the need to lament changes in our global ecosystems. How can we change our hearts when we have, as Bill McKibben observes in his book *The Comforting Whirlwind*, "raised *more* on a pedestal?"[80] In other words, we are comfortable in our resource use, always desiring more; we are distanced from most disease and famine and rarely think about the suffering of our neighbors. Like Job's friends, we believe that "more is better, growth is necessary," and that we deserve to prosper.[81] According to McKibben, Job is about changing paradigms. For Job, it was challenging the paradigm that God punishes the wicked; for us, perhaps we need to challenge the "orthodoxy of more."[82]

How do we challenge comfort? How can we open our eyes to the need for lament? Story can play an important role in waking us up to the need for lament. In this chapter, we have been considering the story of how we have altered the earth

---

[78]Rah, *Prophetic Lament*, 21.
[79]Veldman, *Gospel of Climate Skepticism*, 2.
[80]Bill McKibben, *The Comforting Whirlwind: God, Job and the Scale of Creation* (Cambridge, MA: Cowley, 2005), 8.
[81]McKibben, *Comforting Whirlwind*, 25.
[82]McKibben, *Comforting Whirlwind*, 16.

through our use of resources and our desire for more. How we hear and respond to this story will differ based on the other stories we have heard throughout our lives. Stephen Bouma-Prediger, one of the most important voices in the creation care movement, says that "the kind of person we become depends on the stories with which we identify."[83] In *The Magician's Nephew*, we see this firsthand. Where Digory hears music, Uncle Andrew hears "only a snarl . . . and barkings, growlings, bayings, and howlings."[84] Alan Jacobs, in his book *The Narnian*, explains that Lewis often creates stories with characters who are blind to anything but their own stories.[85] Are we like the characters in Lewis's stories, who are "incapable of careful looking and attentive listening?"[86] Is it possible that, like the dwarves in *The Last Battle* who claim "the dwarfs are for the dwarfs," that we are so comfortable in our pursuit of "more" that we live only for ourselves?[87] When we read favorite authors such as Lewis and Tolkien and observe their love and respect for nature, we may begin to hear the story of responsibility toward creation in a new way. If we can listen well and truly hear the stories of suffering, we may begin to realize that we are called to lament.

How is it possible for the privileged to truly lament for creation when we do not experience firsthand the suffering of creation and our neighbors? Denise Ackermann argues in *After the*

---

[83] Bouma-Prediger, *Earthkeeping and Character*, 15.

[84] C. S. Lewis, *The Magician's Nephew* (New York: Collier, 1955), 126.

[85] Alan Jacobs, *The Narnian: The Life and Imagination of C. S. Lewis* (New York: HarperCollins, 2005).

[86] Jacobs, *Narnian*.

[87] C. S. Lewis, *The Last Battle* (New York: Collier, 1956), 73. It is also worth noting that the dwarfs' inability to view reality without bias leads to their rejection of Aslan's gifts of good food and drink, which they mistake for old hay and muddy water. Thus we are reminded of Tolkien's admonition to use story to help us "recover" clear understanding.

Locusts that there are different ways that each of us react to suffering. Some respond by succumbing to apathy.[88] According to Ackermann, "the purpose of apathy [is] to block awareness of suffering to such an extent that one becomes immune to it."[89] The residents of Hobbiton had become apathetic even as they saw their beloved Shire changing at the hands of Saruman; it took the return of the four hobbits to awaken them to this evil. Even the Ents had become apathetic to the destruction caused by Saruman until the hobbits' presence seemed to rouse them to the suffering of the trees. Others respond to suffering by withdrawing into silence, which could be, according to Ackermann, "respectful, even compassionated" but often "signal[s] withdrawal, not into apathy, but into inactive spectatorship."[90] For still others, suffering can lead to anger or even atheism,[91] and this is precisely why lament is so important. Lament offers us an alternative to believing that God is not interested in suffering. Consider the laments that are offered to us as prayers in the Psalms, or the words of Paul in Romans 8:22: "It is plain to anyone with eyes to see that at the present time all created life groans in a sort of universal travail" (Phillips NT). Have you ever considered that creation itself is groaning in lament as a result of the suffering in our natural world that was caused by the fall? According to Rah, "lament is an act of protest as the lamenter is allowed to express indignation and even outrage about the experience of suffering."[92] Rah explains that "the lamenter talks back to God and ultimately petitions him for help, in the midst of pain. The one who laments can call out to God for help, and in that outcry,

---

[88]Denise Ackermann, *After the Locusts: Letters from a Landscape of Faith* (Grand Rapids: Eerdmans, 2003).

[89]Ackermann, *After the Locusts*, 101.

[90]Ackermann, *After the Locusts*, 101.

[91]Ackermann, *After the Locusts*, 101.

[92]Rah, *Prophetic Lament*, 44.

there is the hope and even the manifestation of praise."[93] Lament is "a language that is honest, does not shirk from naming the unbearable, [and] does not lie down in the face of suffering or walk away from God."[94] Lament offers us a way forward—a way to respond to the suffering of our neighbors and the earth.

Lament prepares us for action, but sometimes it is still difficult to find a way forward. Ackermann alerts us to the warning signs of apathy, silence, and disbelief, but as Dickerson and Evans argue, Tolkien, in his storytelling, provides us with a way to move forward with environmental action through "recognition that inaction results in further harm, abandonment of despair [for] the trust that positive actions have positive consequences, and [ultimately] sufficient care for the created world to do something about the danger."[95] Like Treebeard, we need to break free of apathy and recognize that if we "stay home and do nothing" about environmental harm and climate change, "doom would find us anyway, sooner or later."[96] Evangelicals, especially those in the United States, need to question their reluctance to agree that our climate is changing. While most of the world is concerned about changing climates,[97] our own apathy and inaction are causing further harm and pushing us closer and closer to tipping points that could prevent ecosystem repair.[98]

---

[93]Rah, *Prophetic Lament*, 21.

[94]Ackermann, *After the Locusts*, 112.

[95]Dickerson and Evans, *Ents, Elves, and Eriador*, 122.

[96]Tolkien, *Two Towers*, 486.

[97]Moira Fagan and Christine Huang, "A Look at How People Around the World View Climate Change," Pew Research Center, April 18, 2019, www.pew research.org/fact-tank/2019/04/18/a-look-at-how-people-around-the -world-view-climate-change.

[98]Fred Pearce, "As Climate Change Worsens, a Cascade of Tipping Points Looms," Yale Environment 360, 2019, https://e360.yale.edu/features/as -climate-changes-worsens-a-cascade-of-tipping-points-looms.

But unfortunately, despite scientific consensus regarding climate change, all too often the evangelical community remains silent on this issue.[99] This silence is a type of denial (or disbelief). Thus, even if we do not deny the science of climate change, it is still possible that we deny the implications of climate change by our inaction.[100] Moe-Lobeda would argue that this silence is a form of structural violence and exerts harm on our neighbor.[101] She asserts that the power of this type of violence lies in "its ability to remain invisible or ignored by those who perpetuate it or benefit from it."[102] It would be easy to despair once we realize the harm our silence causes, but if we take action, we can see positive change. We learn this from Merry and Pippin in *The Lord of the Rings*. It would be easy to think that these characters are inconsequential because they are small and often in need of rescue. However, in the short time that they were with Treebeard, he was encouraged to act—"'I will stop it!' he boomed. 'And you shall come with me. You may be able to help me.'"[103] Indeed, they were able to help, as Gandalf explained to the Fellowship: "their coming was like the falling of small stones that starts an avalanche in the mountains. Even as we talk here, I hear the first rumblings. Saruman had best not be caught away from home when the dam bursts!"[104] We should follow the hobbits' example and fight against apathy, silence, and

---

[99]"Scientific Consensus: Earth's Climate Is Warming," NASA, 2020, https://climate.nasa.gov/scientific-consensus/.

[100]Douglas Kaufman, "Caring About Climate Change: An Anabaptist Cruciform Response," *Mennonite Quarterly Review* 94 (2020): 83-102.

[101]Moe-Lobeda, "Climate Change as Climate Debt," 27-49.

[102]Moe-Lobeda, "Climate Change as Climate Debt."

[103]Tolkien, *Two Towers*, 474.

[104]Tolkien, *Two Towers*, 496.

despair and do our part to push back against the "orthodoxy of more."[105]

There is no doubt that lament is difficult when you are not personally experiencing the suffering; however, we are all called to lament the effects of our consumerism on creation. Christian teaching about *imago Dei* reinforces the idea that humans are "set apart" from creation. This tendency to see ourselves as distinct from nature facilitates the overuse of resources and the transformation or destruction of landscapes. These behaviors are justified by an interpretation of Genesis 1:28 to "have dominion over and subdue creation." Understanding Genesis in this manner allows us to rationalize overuse of resources and an economy built on the perceived abundance of "more." There is no doubt that resource use in the developed world is built on the economic value of resources. In fact, the father of American forestry and the first chief of the United States Forest Service, Gifford Pinchot, described the importance of prosperity to his model of resource management when he said, "Without natural resources life itself is impossible. From birth to death, natural resources, transformed for human use, feed, clothe, shelter, and transport us. Upon them we depend for every material necessity, comfort, convenience, and protection in our lives. Without abundant resources prosperity is out of reach."[106] When prosperity is our goal, we end up with landscapes like the one described in C. S. Lewis's poem, "The Future of Forestry," in which trees are gone, landscapes are covered with pavement, and children only know of trees from stories they hear and read:

---

[105]McKibben, *Comforting Whirlwind*, 16.
[106]Gifford Pinchot, *Breaking New Ground* (Washington, DC: Island, 1998), 505.

How will the legend of the age of trees
feel, when the last tree falls in England?[107]

Envisioning a world where there are no trees should make
us change our behaviors so that we are "more than ourselves"
and see the future implications of our consumption.

Accordingly, it seems that the first step must be to help
people recognize that their behavior is linked to global environmental problems. Bill McKibben notes that many church-
goers view creation care as "a luxury you would get to after
you'd dealt with poverty and war."[108] What many do not understand is that war and poverty are directly linked to environmental damage, as many conflicts begin over rights to
resources. Additionally, conflict damages the land and exac-
erbates or perpetuates poverty through decreased access to
resources or displacement of human populations.[109] Environmental damage and climate change are also combining to increase the suffering of vulnerable neighbors by accelerating
factors that lead to drought, famine, and disease emergence.[110]
Sadly, those impacted the most by changing ecosystem services are the least likely to consume the resources that fa-
cilitate these changes. Likewise, those of us who consume the
most are the least likely to experience negative effects, and

---

[107]C. S. Lewis, "The Future of Forestry," 1938, www.poeticous.com/c-s-lewis
/the-future-of-forestry.

[108]Whitney Bauk, "'The Dominant Theological Issue': Environmentalist Bill
McKibben Wants Your Pastor at the Global Climate Strike," *Washington
Post*, September 19, 2019, www.washingtonpost.com/religion/2019/09/19
/dominant-theological-issue-environmentalist-bill-mckibben-wants
-your-pastor-global-climate-strike/.

[109]Colleen Wanner, "War and Environmental Degradation: A Shared History,"
2016, www.earthday.org/war-and-environmental-degradation/.

[110]These ideas were also presented as a part of Wheaton College's TowerTalk
series. See Kristen Page, "The Impact of Creation Care on Public Health,"
Wheaton College TowerTalks, 2018, www.wheaton.edu/academics/faculty
/towertalks/the-impact-of-creation-care-on-public-health/.

we are therefore not likely to be motivated to change this behavior. However, Scripture calls us to lament the suffering of both creation and our neighbor. In this way, lament leads us to action and a shifting of the paradigm of "more." We need to realize that, like Edmund and Turkish delight, we will always want more, yet we will never be satisfied. In spite of very significant challenges, we can fight against apathy, silence, and inaction and begin to alter the behaviors that have resulted in the suffering of our neighbors and a planet in crisis. Just like in Narnia, we may be living in a "winter without a Christmas," but if Christians can lead the way in moving the privileged away from the "orthodoxy of more," together we will be able to make real changes that in turn will loosen the grasp of winter on our global world and bring about the hope of spring once again.

RESPONSE

Noah Toly

Though this second of Dr. Page's three Hansen Lectures is perhaps the most challenging, I consider it a great honor and high calling to respond to this particular chapter. If these three chapters were Page's *Oresteia*, this one would be her *Libation Bearers*; if this series were her *Star Wars*, this lecture would be her *Empire Strikes Back*; if it were her *Lord of the Rings*, this would be her *Two Towers*. In other words, this penultimate chapter would be the long, forbidding middle of her trilogy, and it is from the long, forbidding middle that transformative potential emerges.

I happen to love the middles of many trilogies. I love their *in media res* beginnings and their unresolved ends. They rhyme with a time like ours, where people can see the end—like a trickle of light through the very edge of Mirkwood Forest—but they don't know just how far away it is or whether they'll make it. They rhyme with a time between times—the long stretch between the execution, resurrection, and ascension of Jesus Christ and his coming again in power to finally deliver us from both the terrors of the world and the errors of our ways. They rhyme with a time when the Spirit of

God puts the possibilities of transformation within our reach, but the powers and principalities keep those possibilities just beyond our grasp. They rhyme with a time when the whole creation can glimpse the glorious end but groans through the long, inglorious middle.

It may make some of us uncomfortable to admit it, but without these long, forbidding middles, our stories would be missing something. I don't mean something optional or even merely important, but something essential. Imagine the Chronicles of Narnia with no Stone Table and no winter, as if the children walked through the wardrobe and entered the beautiful Narnian spring. Without the winter, the changing of the seasons loses its meaning. Alternatively, imagine Gandalf telling Bilbo not to worry about returning the ring—you don't have to face a perilous and uncertain journey to the Cracks of Doom in the depths of Orodruin and the Fire Mountain while being beset by foes (and sometimes by friends), all while being hollowed out by temptation. Instead, you just box the ring back up, be sure to include the packing slip and shipping label, and set it out at Bag End for UPS to take back to Mount Doom, where some kind employee will return the ring to inventory and credit your account a few days later. That would not only diminish the length of the account but also change the ultimate point and impact of the story. It would turn the story of Sauron's final defeat into a cheap trick if the triumph could be so certainly accomplished at any time with so little cost. It would not be the against-all-odds triumph of justice and joy in Middle-earth. It would no longer be about the Fellowship's exercise of responsibility and growth in virtue, but it would be about a sort of technical, bureaucratic, or logistical success. The story of Frodo and Sam standing at a customer service desk saying they'd like to return a ring would be, at best, transactional rather than transformational.

Thankfully, our stories are darker than that, because responding rightly to the world around us—including to key challenges in sustainability, public health, and environmental justice—depends on nothing less than transformation. The history of these areas of study, policymaking, and activism is dotted with the hard lesson that knowledge and technique are never enough. Rather, some deeper transformation is required. Take, for example, the case of climate change, where it's certainly not for lack of data or scientific consensus that we have made such little progress toward a climate-stable future. What we need is a transformation of our perspective, desire, judgment, and hope.

That's what all these long, forbidding middles do. Whether vicariously or intimately, they not only enrich or enhance our stories, imbuing them with greater meaning or heightening the tension so that the end is somehow more intense or satisfying, but they transform our way of seeing the world. They are, to adapt a term from the late Cardinal Paulo Evaristo Arns, archbishop of São Paulo, *pedagogies of subversion*. According to Cardinal Arns, "To subvert means to turn a situation around and look at it from the other side. That is, from the side of the people who have to die so that the system can go on." The experiences, stories, and practices that instruct us in this way are, as Dietrich Bonhoeffer writes, experiences of incomparable value that teach us to see "from the perspective of the outcasts, the suspects, the maltreated, the powerless, the oppressed and the reviled . . . from the perspective of the suffering."[1] They not

---

[1] Dietrich Bonhoeffer, *Letters and Papers from Prison*, vol. 8 of *Dietrich Bonhoeffer Works*, ed. Victoria J. Barnett and Barbara Wojhoski (Minneapolis, MN: Fortress Press, 2010), 52.

only teach us to *see*, but they transform our desires and judgments. Adding, as Page does, practices of lament—admitting and mourning that things are not the way they're supposed to be—quickly crowds out our desires for the status quo or judgments that everything is fine just as it is.

And desire and judgment transformed by lament through the long, forbidding middle teaches us to hope truly. It's the bondage in Egypt that teaches us the meaning of freedom in the promised land. It's the lengthy exile that instructs us in proper longing, not the false prophecy of a two-year return. It's only you who hunger now who shall be satisfied, and only you who weep now who shall laugh. Whatever else we take those beatitudes to mean, it is fair to wonder whether the blessed are those whose "divine dissatisfaction," to borrow a term from Dr. Martin Luther King Jr.,[2] is so deep that only the kingdom of God can possibly match it. When it does, the satisfaction is incomparably great.

Writing from prison to his friend Eberhard Bethge on the second Sunday of Advent, 1943, Bonhoeffer had something to say about "penultimate words." He wrote,

> Only when one knows that the name of God may not be uttered may one sometimes speak the name of Jesus Christ. Only when one loves life and the earth so much that without it everything seems to be lost and at its end may one believe in the resurrection of the dead and a new world. Only when one accepts the law of God as binding for oneself may one perhaps sometimes speak of grace. And only when the wrath and vengeance of God

---

[2]Martin Luther King, Jr., "Where Do We Go From Here?," in *A Testament of Hope: The Essential Writings and Speeches of Martin Luther King, Jr.*, ed. James M. Washington (San Francisco: HarperCollins, 1986), 251.

against God's enemies are allowed to stand can something of forgiveness and the love of enemies touch our hearts. . . . One can and must not speak the ultimate word prior to the penultimate.

Again: "Only when one loves life and the earth so much that without it everything seems to be lost and at its end may one believe in the resurrection of the dead and a new world. . . . *One can and must not speak the ultimate word prior to the penultimate.*"[3]

We may look forward to Page's third and final lecture on regaining wonder and joining the chorus, but we must recognize that we are looking at it from a long way off. I would say that we can't wait for it, but we *must* wait. Don't shortchange the transformation by hurrying on to the end. Instead, put the book down and let this chapter sink in. The ultimate word must not come before the penultimate.

---

[3]Bonhoeffer, *Letters and Papers*, 213.

ASK THE ANIMALS
TO TEACH YOU

How to Regain Wonder and Rejoin the Chorus

We share the earth with between eight million and one trillion other species, most of which have not even been discovered because they are too small or live in habitats that are difficult to access.[1] When I think about all of the amazing creatures we share the planet with, I am overwhelmed with curiosity! I am driven to want to know more. Where do they live? What do they look like? How do they interact with other species? I have always been this way. It is how I was created, and learning about the diversity of life on the planet brings me great joy. I think that it also brings God joy because we read in Job, "But ask the animals, and they will teach you, or the birds in the sky, and they will tell you; or speak to the earth, and it will teach you, or let the fish in the sea inform you.

[1]Michael Benton, "Origins of Biodiversity," PLOS *Biology* 14, no. 11 (2016): e2000742; Gerardo Ceballos et al., "Accelerated Modern Human-Induced Species Losses: Entering the Sixth Mass Extinction," *Science* (2015): 1e1400253; Kenneth Locey and Jay Lennon, "Scaling Laws Predict Global Microbial Diversity," PNAS 113, no. 21 (2016): 5970-75.

Which of all these does not know that the hand of the Lord
has done this?" (Job 12:7-9).

I understand this to mean that the diversity of life that God
created is important and worth knowing. Indeed, our first
God-given responsibilities were to care for the garden and to
name the animals (Gen 2:15, 19). These are tasks that imply
that we should learn about creation, as naming implies
knowing—and to care properly for a garden requires
knowledge of what is needed by the life within. This passage
in Job also suggests that we can learn something about God
as we develop knowledge of his creation. Here is how Dorothy
Sayers described the divine knowledge that comes from observing nature and how this practice helps us learn more
about creation by understanding the Creator: "Why should
God, if there is a God, create anything, at any time, of any kind
at all? . . . The Church asserts that there is a Mind which made
the universe, that He made it because He is the sort of Mind
that takes pleasure in creation, and that if we want to know
what the Mind of the Creator is, we must look at Christ."[2]

In order to respond to the call to care for and know creation, we must also address the issue that the created diversity of our natural world is declining. Indeed, we are currently living through the sixth mass extinction of organisms
in the history of the Earth. Thanks to the fossil record, we
know that extinctions can occur even without human influence, and this background extinction rate is estimated to
be approximately two extinctions per ten thousand species
per one hundred years.[3] However, since the 1800s, extinction

---

[2]Dorothy Sayers, "The Triumph of Easter," in *Creed or Chaos?* (London: Methuen,
1947), 9-10.
[3]Ceballos et al., "Accelerated Modern Human-Induced Species Losses."

rates have increased dramatically and have been recorded as high as eight to one hundred times these background rates of extinction. Using the fossil record as well as more recent evidence compiled by the International Union for Conservation of Nature (IUCN), 617 vertebrate species have either been documented as extinct, have become extinct in the wild, or have been listed as possibly extinct since 1500.[4] According to the IUCN, there are over 35,000 species in danger of extinction right now.[5] In my lifetime we have seen the extinction of the Caspian tiger, the Guam flying fox, the Siamese flat-barbelled catfish, the Yunnan lake newt, the golden toad, the rotund rocksnail, the Pyrenean ibex, the Pinta giant tortoise, and the West African black rhinoceros.[6]

When I think about how quickly we are losing species, I feel frantic! I feel a deep despair that I have not known more about what God has created before his creatures were gone. I realize that this seems dramatic, and that most people do not react this way. But most people have not seen the amazing things I have seen in my life, and most people do not realize that we are losing these species as a result of our pursuit of comfortable lives. Consider what we read in Genesis as paraphrased from *The Message*:

> God spoke: "Swarm, Ocean, with fish and all sea life!
> Birds, fly through the sky over Earth!"
> God created the huge whales,
> all the swarm of life in the waters,
> And every kind and species of flying birds.

---

[4]Locey and Lennon, "Scaling Laws."
[5]IUCN, "The IUCN Red List of Threatened Species," 2020, www.iucnredlist.org.
[6]Helle Abelvik-Lawson, "18 Animals That Became Extinct in the Last Century," Greenpeace, September 10, 2020, www.greenpeace.org.uk/news/18-animals-that-went-extinct-in-the-last-century/.

God saw that it was good . . .
God spoke: "Earth, generate life! Every sort and kind:
cattle and reptiles and wild animals—all kinds."
And there it was:
wild animals of every kind,
Cattle of all kinds, every sort of reptile and bug.
God saw that it was good. (Gen 1:20-25)

Reading the biblical story of God's creative and imaginative design for the diversity of the animal kingdom, how can we not mourn the loss of all that he so intentionally created? Each time a species is lost, we lose an aspect of the Creator's intentional design, a gift that he has given us for a specific purpose. We should never forget the divine declaration made by the Creator as he observed the myriad animals he had just created: "And God saw that it was good."

I have always been captivated by animals. As a young child, I spent hours playing with my collection of plastic animals, creating and enacting stories that must have been something similar to how I now envision the reconciled earth: lions, lambs, zebras, bears, eagles, and giraffes all playing together joyfully in the imagined landscapes of my bedroom. I also spent hours playing outside, whether catching crayfish and tadpoles in the creek or walking through the woods behind our house. I can remember sitting under the large branches of an old magnolia. This felt like my own world, a place that fostered my curiosity and creativity. My curious nature was fostered by my family—especially my grandmother, Nanny, who spent time with me in her garden teaching me about the plants and the birds visiting her feeder. My mother also taught me to pay attention to the beautiful details around me, especially while watching birds or taking long walks at the beach, where we picked up beautiful shells and amazing fossils that sparked my imagination about creatures of the past.

When I began my undergraduate studies in biology, my very first course was zoology with a professor who would become one of my most important mentors, Dr. Bill Teska. I loved this course. I loved Dr. Teska's excitement when he showed us something amazing about animals, and I loved the delight he took in our learning. I took every course I could with this passionately curious man! The summer after my freshman year, I spent a month traveling in a van with Dr. Teska and twelve other students, camping our way across the United States. We hiked the Grand Canyon from rim to rim, and we used black light to catch scorpions and spotlights to catch kangaroo rats. We birded from Zion to the Olympic Peninsula and then saw our first bears in Glacier National Park. We set small mammal traps every night so that we could see the diversity of rodents and insectivores across North America. We always released everything unharmed, and I was absolutely in love with learning! To hold an animal in your hands, to learn something amazing about the way they use their habitat and interact with other species does something to me! It makes me wonder! It makes the passage in Job seem possible. "But ask the animals . . ." Yes, ask the animals!

In my senior year of college, I studied ecology in Costa Rica and Ecuador. I saw and learned so many amazing things during this time, but perhaps the most amazing thing that happened was on the small boat we lived on for two weeks in the Galapagos Islands. The captain told us there were whales in the area, and I went up to the deck just in time to see a humpback whale swimming straight at the side of our boat. At what seemed to be the very last second, the whale turned on its side and looked up at me. Looking into the eye of that whale was one of the most wondrous moments of my life. I felt so many things in that moment. I certainly experienced awe and joy. I was so curious about this whale! I had so many questions. In this moment, the combination of all these feelings made me feel compelled to act—to learn, to change, and even to protect. This experience only lasted a minute, if that, but it had a profound effect on me. Have you ever experienced something like this?

My encounter with the whale was wondrous! But I have had other experiences that have stirred similar feelings, and for me they often happen while camping or hiking or wandering beyond a trail. Not everyone is as compelled as I am to experience creation in such a way, nor is everyone physically able to hike through natural landscapes due to reasons of health or opportunity. Nevertheless, I do believe that everyone can pursue wonder through making the effort to spend time in fictional landscapes. I remember the first time I heard Aslan's song in *The Magician's Nephew*:

> In the darkness something was happening at last. A voice had begun to sing. . . . the voice was suddenly joined by other voices; . . . the blackness overhead, all at once, was blazing with stars. . . . The lion was pacing to and fro about that empty land and singing his new song. . . . And as he walked and sang the valley grew green with grass. It spread out from the Lion like a pool. It ran up the sides of the little hills like a wave. In a few minutes it was creeping up the lower slopes of the distant mountains, making that young world every moment softer.[7]

---

[7]C. S. Lewis, *The Magician's Nephew* (New York: Collier, 1955), 98-99.

I was transported back to beautiful landscapes that I have experienced in my life. As I listen to the song, I revisit in my mind places where I have experienced deep joy and places that have left me with many, many more questions than answers.

Lani Shiota, a social psychologist who studies emotion, describes "wonder as that moment when our minds are trying to stretch, to take in and comprehend whatever it is that's before us."[8] C. S. Lewis invites us to wonder through Aslan's song, and we are encouraged to comprehend the beauty of what we are seeing and hearing. Stephen Bouma-Prediger, in his book, *Earthkeeping and Character*, suggests that wonder is a virtue in which we "stand in rapt attention and amazement in the presence of something awe-inspiring, mysterious or novel."[9] As the story of the creation of Narnia unfolds, we are amazed to see everything from moles, frogs, elephants, and even a lamp post emerge. As we contemplate the cacophony of these beginnings, something stirs in us. Wonder is more than just a feeling of reverence; it is an action—it is the moment we start engaging our curiosity by asking new questions.[10] Wonder moves us to action! "The Lion was singing still. But now the song had once more changed. It was more like what we should call a tune, but it was also far wilder. It made you want to run and jump and climb. It made you want to shout."[11]

If we understand wonder as virtue, as suggested by Bouma-Prediger, "we exhibit the virtue when we have the cultivated capability to stand in grateful amazement at what God has

---

[8]Steve Paulson et al., "Unpacking Wonder: From Curiosity to Comprehension," *Annals of the New York Academy of Sciences* (2020): 1-20.
[9]Steven Bouma-Prediger, *Earthkeeping and Character: Exploring a Christian Ecological Virtue Ethic* (Grand Rapids, MI: Baker Academic, 2020), 43.
[10]Paulson et al., "Unpacking Wonder."
[11]Lewis, *Magician's Nephew*, 113.

made and is remaking."[12] I believe this gratitude for the wonder and beauty of nature should then lead us to the desire to act through stewardship of creation. Philosopher Martyn Evans seems to agree that wonder should move us to action and suggests that wonder is a "transfiguring encounter" that results in an "altered, compellingly intensified attention to something that we immediately acknowledge as somehow important."[13] As readers experience wonder through their experiences in fictional landscapes, could they be changed in a way that will impact their interaction with actual landscapes? I believe so. For me, I am moved to action when I hear Aslan's song. For the moment, I am *in* Narnia, and I respond to Aslan's request "to be": "Narnia, Narnia, Narnia, awake. Love. Think. Speak. Be walking trees. Be talking beasts. Be divine waters."[14]

Spending time in fictional landscapes also reminds us of the many amazing aspects of creation that should cause us to wonder. These reminders are necessary as we are easily distracted by the obligations of our lives. We need to be more like children and approach the world with a wide-eyed sense of wonder and endless questions. The distractions of adult lives make us forget the joy and excitement of learning something new about creation and being amazed. Rachel Carson, the pioneering conservation ecologist, noted, "It is our misfortune that for most of us that clear-eyed vision, that true instinct for what is beautiful and awe-inspiring, is dimmed even lost before we reach adulthood."[15] This loss of a sense of wonder has important implications for our lives. We are more likely to experience

[12]Bouma-Prediger, *Earthkeeping and Character*, 43.
[13]H. M. Evans, "Wonder and the Clinical Encounter," *Theoretical Medical Bioethics* 33 (2012): 127.
[14]Lewis, *Magician's Nephew*, 116.
[15]Rachel Carson, *Sense of Wonder* (New York: Open Road, 1965), 84-85.

wonder when we are surrounded by nature, yet most of us spend more of our lives indoors than out, and it is estimated that the average person will spend at least 90 percent of their life indoors.[16] Recently, much attention has been given to the health benefits of spending time in nature. Many studies demonstrate how time spent in nature can improve our health by decreasing stress, heart rate, blood pressure, and cholesterol.[17] Spending time in nature can decrease our risk for Type 2 diabetes, pre-term births, and even death due to cardiovascular disease.[18] Walking on forested paths has been demonstrated to reduce perceptions of stress,[19] and spending extended time in forests can reduce physiological markers of stress and increase feelings of comfort and refreshment.[20] Interactions with nature can improve mood for those with depression,[21] shorten time for recovery from surgery and illness,[22] improve self-discipline,[23] and even improve cognition.[24] Surprisingly, you do not even need to be physically in nature to receive the health benefits. Much of the research shows that simply seeing photographs of

[16]Gary Evans and Janetta McCoy, "When Buildings Don't Work: The Role of Architecture in Human Health," *Journal of Psychology* 18 (1998): 85-94.

[17]Caoimhe Twohig-Bennett and Andy Jones, "The Health Benefits of the Great Outdoors: A Systematic Review and Meta-Analysis of Greenspace Exposure and Health Outcomes," *Environmental Research* 166 (2018): 628-37.

[18]Twohig-Bennett and Jones, "Health Benefits of the Great Outdoors."

[19]Masahiro Toda et al., "Effects of Woodland Walking on Salivary Stress Markers Cortisol and Chromogranin A," *Complementary Therapies* 21 (2013): 29-34.

[20]J. Lee et al., "Effect of Forest Bathing on Physiological Responses in Young Japanese Male Subjects," *Public Health* 125 (2013): 93-100.

[21]Marc Berman et al., "Interacting with Nature Improves Cognition and Affects for Individuals with Depression," *Journal of Affective Disorders* 140 (2012): 300-305.

[22]Roger Ulrich, "View Through a Window May Influence Recovery from Surgery," *Science* 224 (1984): 420-21.

[23]Andrea Taylor, Frances Kuo, and William Sullivan, "Views of Nature and Self-discipline: Evidence from Inner-City Children," *Journal of Environmental Psychology* 22 (2002): 49-63.

[24]Marc Berman, John Jonides, and Stephen Kaplan, "The Cognitive Benefits of Interacting with Nature," *Psychological Science* 19 (2008): 1207-12.

natural scenes has benefits for health.[25] Scientists have also linked improved health with reading fiction,[26] so it is possible that spending time in fictional landscapes, especially ones with descriptions of nature, has health benefits. Reading fiction is important, and "our interactions with fictional narratives should not be viewed as frivolous; stories have the power to change our beliefs about the real world."[27]

Tolkien seems to recognize the health benefits of nature and literature. In "On Fairy-Stories," he describes himself as a "wandering explorer . . . in the land, full of wonder but not of information."[28] He was speaking of wandering in and exploring fairy stories, of course; however, anyone who spends time in the fictional landscapes that Tolkien has created has likewise been given the opportunity to explore details along his fictional paths and wonder at the beauty of the view. Tolkien's description of interactions between the Fellowship and the land through which they journey reflects the way Tolkien himself moved through landscapes. George Sayer describes walking with Tolkien as an exercise in interruptions, as Tolkien would frequently stop or slow down in order "to look at the trees, flowers, birds, and insects that [were] passed."[29] Because I also walk in this exploratory way—stopping to listen

---

[25]Taylor Hartig et al., "Nature and Health," *Annual Review of Public Health* 35 (2014): 207-28.

[26]Denise Rizzolo et al., "Stress Management Strategies for Students: The Immediate Effects of Yoga, Humor and Reading on Stress," *Journal of College Teaching and Learning* 6 (2009): 79-88; Ayni Bavishi, Martin Slade, and Becca Levy, "The Survival Advantage of Reading Books," *Innovation in Aging* 1 (2017): 477.

[27]R. A. Mar, "The Neuropsychology of Narrative: Story Comprehension, Story Production and Their Interrelation," *Neuropsycologia* 42 (2004): 1414.

[28]Verlyn Flieger and Douglas A. Anderson, eds., *Tolkien on Fairy-Stories* (London: HarperCollins, 2008).

[29]G. Sayer, "Recollections of J. R. R. Tolkien," in *Tolkien: A Celebration— Collected Writings on a Literary Legacy*, ed. Joseph Pearce, 1-16 (San Francisco: Ignatius, 1999).

to a bird song, to take a picture or to look at something small and often colorful along the trail, I was not surprised to learn that Tolkien shares my slow "reading" of nature. We can tell that he paid attention to the smallest details on his hikes because when we read of the journeys in *The Lord of the Rings*, he describes the smallest details along the journey of the Fellowship. In other words, the time Tolkien spent in nature influenced his understanding of recovery and escape in fairy stories, and this resulted in his ability as an author to use this concept to help both characters within the story as well as readers to benefit from nature and recover. There are many examples in *The Lord of the Rings* in which characters often are restored over time by wondering at the beauty of nature surrounding them. In *The Fellowship of the Ring*, as the Fellowship leaves Moria—grieving the loss of Gandalf, injured, and hungry—they make several restorative stops.

> They stooped over the dark water. At first they could see nothing. Then slowly they saw the forms of the encircling mountains mirrored in a profound blue, and the peaks were like plumes of white flame above them; beyond there was a space of sky. There like jewels sunk in the deep shone glinting stars, though sunlight was in the sky above. . . . "What did you see?" said Pippin to Sam, but Sam was too deep in thought to answer. . . .[30]
>
> Soon afterwards they came upon another stream that ran down from the west, and joined its bubbling water with the hurrying Silverlode. Together they plunged over a fall of green-hued stone, and foamed down into a dell. About it stood fir-trees, short and bent, and its sides

---

[30]J. R. R. Tolkien, *The Lord of the Rings: The Fellowship of the Ring* (London: HarperCollins, 2012), 334.

were steep and clothed with harts-tongue and shrubs of whortle-berry. At the bottom there was a level space through which the stream flowed noisily over shining pebbles. Here they rested.[31]

These stops along the journey to Lothlórien seemed to restore the Fellowship. And by the time the Fellowship first sees Lothlórien, we also see their recovery as they experience the "return and renewal of health."[32]

"There lie the woods of Lothlórien!" said Legolas. "That is the fairest of all the dwellings of my people. There are no trees like the trees of that land. For in the autumn their leaves fall not, but turn to gold. Not till the spring comes and the new green opens do they fall, and then the boughs are laden with yellow flowers; and the floor of the wood is golden, and golden is the roof, and its pillars are of silver, for the bark of the trees is smooth and grey."[33]

Legolas's description of Lothlórien is a remembrance of wonder. As the Fellowship is anxious about their journey, the joint anticipation of this beauty seems to strengthen them and provide a glimpse of needed hope. However, when they reach Lothlórien, they must begin their journey by being blindfolded in solidarity with Gimli the Dwarf, who is only permitted to enter the Elven Realm if his eyes are covered. Thus, at first, they do not experience the beauty with their eyes. Rather, they experienced the beauty and wonder of the place with their other senses.

---

[31]Tolkien, *Fellowship of the Ring*, 335.
[32]Flieger and Anderson, *Tolkien on Fairy-Stories*.
[33]Tolkien, *Fellowship of the Ring*, 335.

They felt the ground beneath their feet smooth and soft, and after a while they walked more freely, without fear of hurt or fall. Being deprived of sight, Frodo found his hearing and other senses sharpened. He could smell the trees and the trodden grass. He could hear many different notes in the rustle of the leaves overhead, the river murmuring away on his right, and the thin clear voices of birds high in the sky.[34]

When the blindfolds are finally removed, Frodo wonders at what he sees:

Frodo looked up and caught his breath. They were standing in an open space. . . . mallorn-trees of great height, still arrayed in pale gold. . . . At the feet of the trees, and all about the green hillsides the grass was studded with small golden flowers shaped like stars. . . . Here ever bloom the winter flowers in the unfading grass: the yellow elanor, and the pale niphredil. Here we will stay awhile . . . The others cast themselves down upon the fragrant grass, but Frodo stood awhile still lost in wonder.[35]

When we are in nature, we are surrounded by interesting stimuli that allow our attention to be released from the need to focus intently on work or specific tasks.[36] In this more relaxed state, we are more likely to experience nature in ways that evoke wonder and awe, and we are encouraged to engage with aspects of creation we may not have had interest in before. Lani Shiota explains the benefit of awe:

---

[34]Tolkien, *Fellowship of the Ring*, 349.
[35]Tolkien, *Fellowship of the Ring*, 350.
[36]Marcus Hedblom et al., "Reduction of Physiological Stress by Urban Green Space in a Multisensory Virtual Experiment," *Scientific Reports* 9 (2019): 10113.

More than any other species on Earth, humans are profoundly dependent on knowledge. . . . that allows us to
map our environment, remember the past, and predict
the outcomes of future actions, all within the scope of
human imagination. The emotion we call awe—our capacity for deep pleasure in facing the incredible and
trying to take it all in—may reflect a basic need to understand the world in which we live.[37]

Awe is the first step to developing the virtue of ecological
wonder, which is so important to enabling our response to the
call to be stewards of creation.[38]

Experiencing wonder can be transformational, and this can
have important implications for Christians. If we believe what
we read in Romans that "God's invisible qualities—his eternal
power and divine nature—have been clearly seen, being understood from what has been made, so that men are without
excuse" (Rom 1:20), then spending more time in creation may
open our minds and help us understand more about our
Creator and transform us so that we develop a virtue of
wonder and gratitude. Paying close attention to the created
world around us helps us develop an attitude of openness to
wonder. Robin Wall Kimmerer, an indigenous botanist and
nature writer, agrees and explains that "attention is a doorway
to gratitude, a doorway to wonder, a doorway to reciprocity."[39]
I often refer to the practice of deliberate attention to nature

---

[37]Michelle Laini Shiota, "How Awe Sharpens Our Brains," *Greater Good Magazine*, 2016, https://greatergood.berkeley.edu/article/item/how_awe
_sharpens_our_brains.

[38]Bouma-Prediger, *Earthkeeping and Character*.

[39]Robin Wall Kimmerer, "The Intelligence of Plants," *On Being*, podcast hosted
by Krista Tippet, February 25, 2016, https://onbeing.org/programs/robin
-wall-kimmerer-the-intelligence-of-plants-2022/.

as "reading landscapes," and in many of my courses I try to help my students develop this discipline. I typically ask my students to spend time closely observing the details of a place—paying close attention to everything from what they see and hear to what they smell and feel. In practicing this discipline, I hope that they learn that when we slow our lives and take time to engage in a close reading of creation,[40] we are then able to turn our focus and wonder toward the Creator. In this way, spending time exercising deep attention to creation can be similar to a religious reading of written texts because it facilitates a relationship between ourselves, the reader, and God, the Author or Creator.[41] What we learn from this type of reading, whether of landscapes or of religious works, should "yield meaning, suggestions (or imperatives) for action, matter for aesthetic wonder, and much more."[42]

One important element of wonder is that it helps us develop "an attitude of openness or receptivity that leads from a preoccupation with self into a search for meaning beyond oneself."[43] C. S. Lewis describes such a transformation in *The Voyage of the Dawn Treader*. In this story, Lucy and Edmund return to Narnia with their cousin Eustace Scrubb. Eustace is an unpleasant, friendless, and "beastly" boy who has never believed the tales of Narnia told by his cousins. Finding himself in Narnia on the Dawn Treader, he still fails to see the wonders of Narnia due to his own

---

[40]Alan Jacobs, "The Attentive Reader (and Other Mythical Beasts)," November 24, 2014, http://blog.ayjay.org/uncategorized/attentivereader/.

[41]L. Kristen Page, "Reading the Landscape: Religious Reading and Contemplation," Wheaton College Center for Applied Christian Ethics, May 2015, www.wheaton.edu/academics/academic-centers/center-for-applied-christian-ethics/cace-faculty-article-series/2015-16-cace-faculty-article-series/kristen-page/.

[42]Paul Griffiths, *Religious Reading: The Place of Reading in the Practice of Religion* (New York: Oxford University Press, 1999), 41.

[43]Kathleen Moore, "The Truth of the Barnacles: Rachel Carson and the Moral Significance of Wonder," *Environmental Ethics* 27 (2005): 269.

self-absorption, and he further alienates himself from his cousins and the crew of the Dawn Treader with constant complaining and grumbling. Following a harrowing storm, the ship in need of repair is anchored off a mountainous island. Before he can be asked to help with the work, Eustace leaves the group and begins to explore the island on his own. It is at this point he has an encounter that will begin his transformation. If wonder occurs as a result of something surprising or unexpected and stretches our mind to consider the world in a new way,[44] then certainly an encounter with a dragon could be considered a wonderful (even if terrible) experience. Not only did Eustace see a dragon, he became a dragon, and through this experience realizes that he had always been a beast within: "He realized that he was a monster cut off from the whole human race. An appalling loneliness came over him. He began to see that the others had not really been fiends at all. He began to wonder if he himself had been such a nice person as he had always supposed."[45]

The loneliness is a necessary part of the transformation that Eustace makes, and ecologist Rachel Carson would argue that "there is loneliness in a sense of wonder too, what she called 'a sense of lonely distances,' as we feel our isolation from what is profoundly apart. Loneliness turns to yearning, a kind of love, an overpowering attraction to something beautiful."[46] Perhaps, in this way, wonder draws us to God. Through an experience with wonder, Eustace is isolated from his fellow travelers, experiences this lonely distance, and discovers something about himself. Ultimately his "character

---

[44]Laura-Lee Kearns, "Subjects of Wonder: Toward an Aesthetics, Ethics, and Pedagogy of Wonder," *Journal of Aesthetic Education* 49 (2015): 98-119; Paulson et al., "Unpacking Wonder," 1-20.

[45]C. S. Lewis, *The Voyage of the Dawn Treader* (New York: Collier, 1952), 76.

[46]Moore, "Truth of the Barnacles," 268.

[was] rather improved by becoming a dragon."[47] The most important part of his transformation required Aslan's intervention. Eustace first had to wonder at his ability to understand a lion: "I looked up and saw the very last thing I expected: a huge lion coming slowly toward me. . . . it came close up to me and looked straight into my eyes. And I shut my eyes tight. But that wasn't any good because it told me to follow it."[48] Then he had to allow Aslan to "undress him": "The very first tear he made was so deep that I thought it had gone right into my heart. And when he began pulling the skin off, it hurt worse than anything I've ever felt."[49] Aslan's work uncovered more of the true character of Eustace. He was no longer a dragon; however, his transformation was still a work in progress: "It would be nice, and fairly true, to say that 'from that time forth Eustace was a different boy.' To be strictly accurate, he began to be a different boy. He had relapses. There were still many days when he could be very tiresome. But most of those I shall not notice. The cure had begun."[50]

This is how wonder is for us as well. Wonder begins the transformative work. As we pay close attention to nature, we are more likely to notice wonderful aspects of God's amazing creation. We will learn more about the Creator and begin a transformation toward stewardship based on a virtue of ecological wonder.

Wonder appears to be an important part of recognizing our role as stewards and changing our hearts toward the way we interact with creation. How, then, do we cultivate a sense of wonder? Caspar Henderson explains that "wonder opens up

---

[47]Lewis, *Voyage of the Dawn Treader*, 83.
[48]Lewis, *Voyage of the Dawn Treader*, 88.
[49]Lewis, *Voyage of the Dawn Treader*, 90.
[50]Lewis, *Voyage of the Dawn Treader*, 93.

new possibilities. . . . It can feel like the apprehension of something bigger and better of which we are momentarily a part. It can feel like discovery, or at least the first step on a journey towards one. And it can feel like return or recovery—a sense that something is being put right."[51] Tolkien describes this feeling of "return or recovery" as being crucial to transformation and an important goal of his stories.[52] In Tolkien's short story "Leaf by Niggle," we see such a transformation. Niggle is an artist consumed with the desire to paint a wondrous scene that he holds in his imagination, yet he is often interrupted by his neighbor, Parrish, a gardener. Before he is able to complete his painting, he is required to depart on a journey, leaving his painting unfinished. As he was unprepared for the journey, he is forced to spend his days working in unskilled labor and in isolation. While he is gone, the painting is used to repair a roof, and the beautiful creation seems lost. After a time of isolation and separation from his art, Niggle is released and sent away for "gentle treatment" in a forest. When he arrives, Niggle recognizes the forest as the realization of the scene he was painting. In this place, he is reunited with his former neighbor, Parish, and working together they plant gardens and tend the land until it is transformed into the landscape of Niggle's imagination. Like Niggle, we may hold an idea in our imaginations of what is wonderful about this world in which we live. However, we are living in a world experiencing rapid transformations that are actually diminishing the wonder of God's creation. We are losing biodiversity at an alarming rate; climate is changing, and we find ourselves at a tipping point where ecosystems are failing. We

[51]Caspar Henderson, *A New Map of Wonders* (London: Granta, 2017), 24.
[52]Flieger and Anderson, *Tolkien on Fairy-Stories*.

are at a place where we must make a collective decision to change our behaviors and transform our hearts in order to save our natural world from unfolding deterioration. We must rediscover the wonder of creation, work together to bring back a healthy, natural world, and allow the wonder to move us to rejoin creation's chorus.

If you have ever spent any time in nature with a child, then you know how passionately observant and curious they are. They are closer to the ground, and they notice details that adults often trample. Through their observations and questions, they help adults regain a sense of wonder about everyday occurrences like sunsets. I'll never forget when my daughter, Wren, called me into the room exclaiming, "Mommy, hurry, the sunset has all of my favorite colors!" Children are very good at wondering and recognizing the wonderful things about creation. Wonder comes naturally to them; they just need to be given the opportunity to experience nature. My family loves to camp and hike, and we especially love to be places where we can see birds during the day and stars at night. Several summers ago, Wren, my mom, and I spent an incredible evening star gazing in Canyonland National Park in Utah. We found an out-of-the-way pulloff and parked the car before sunset. While we waited on the sunset, Wren enjoyed running around on the rocks and watching the ravens catch the uplifts from the canyons below. She also had her cello with her, and as she played, we all delighted in the response of the ravens. Several flew in close and appeared to be listening. Wren was improvising—playing the music that came to her as she experienced creation. As my mom and I watched, we wondered at the musical conversation between a girl, her cello, and the ravens. As if this were not wonderful enough, several hours later, we sat in amazement as the Milky Way appeared, and we were lost in the majesty of the heavens. As wonderful as this night in Utah was, you do not have to go far from home to experience wonder. Spending a year close to home, thanks to the Covid-19 pandemic, my family and I have discovered that the birds in our backyard are just as curious about the cello as the ravens were. We watch in wonder as the birds fly in while Wren plays—everything from flycatchers and sparrows to redstarts and orioles have come to visit. The cardinals always fly in to listen, and that just amazes me! We have discovered new places to walk together and explore; we especially enjoy sunset hikes. And thanks to the approach of the comet Neowise in the summer of 2020 and our strong desire to see it, we have also discovered that we don't have to go too far from our suburban home to find skies dark enough to wonder at the night sky.

I am blessed to be part of a church that worships with children through wonder. We encourage creativity and questions, and we share in wondering as we experience together the stories of the life of Christ. When you wonder while listening in this way, you ask open-ended questions—questions that have no firm answer. When we (like our children) engage with Scripture by wondering, we foster a deep curiosity in them as well as in ourselves. This leaves each one of us desiring to know more about God and his creation. In the Anglican tradition, when a child is baptized, the church is asked if those who witness the vows will do all in their power to support these persons in their life in Christ. When we affirm this, we are promising so much more than being Sunday school teachers. We are promising to journey with the child in their faith. What does this look like? How do we truly journey with a child in their faith? An important part of this journey is embracing the wondrous approach children take to learning. Ask questions with them about the natural world, and then work to protect what they recognize as wonderful in God's creation. I'm sure if we journeyed in our own lives of faith as the child does, we would all have pockets full of rocks and acorns, and we would have many questions! In Matthew 19:14, Jesus says, "Let the little children come to me . . . for the kingdom of heaven belongs to such as these." The gift to us is that if we truly journey in faith with children, we also may grow to understand more about wonder and the incredibly beautiful and fascinating created world where we live, work, and play.

Often we experience wonder in situations where we can recognize the scale of creation. When we are at the summit of a mountain looking out at the horizon, or standing on the edge of a canyon, or even hiking among ancient trees, we might wonder at how small we feel in relation to the expanses of creation. Steven Bouma-Prediger explains that experiencing wonder in this way helps us to develop "the virtue of ecological humility . . . the settled disposition to act in such a way that we know our place and fit harmoniously into it."[53] He explains that if we are "ecologically humble, we acknowledge that we are finite and fallen and thus have an honest and accurate estimation of our abilities and capacities."[54] Norman Wirzba explains that "humility is central to human life because it is through a humble attitude that we must fully approximate our true condition as creatures," and that by developing humility we can experience "the heart of an embodied

---

[53]Bouma-Prediger, *Earthkeeping and Character*, 45.
[54]Bouma-Prediger, *Earthkeeping and Character*.

and spiritual life that is true to the world as a place of belonging and responsibility."[55] Thus, the virtues of ecological wonder and humility that are necessary for us to be dedicated stewards of God's creation are also necessary virtues for our own spiritual development.

Humility joined with wonder is an essential response to the current ecological crisis. We must humbly recognize that when we use technology to continue extracting resources, we can actually damage ecosystem services beyond their capacity to repair. However, with all our technology and modern mechanization, it can be quite difficult to see ourselves in this more humble position. It's often only when we face extreme storms, drought, or other climate catastrophes that we begin to perceive our lack of strength to meaningfully influence or contain the power of nature. Thus it is critical that we exercise wonder in order to help us move toward the humility required to understand our place and role in creation. In the Psalms, David wonders at the beauty of creation and seems to even wonder about his place in creation—that despite the scale of creation, God our Creator knows him and cares for him:

> When I consider your heavens,
> the work of your fingers,
> the moon and the stars,
> which you have set in place,
> what is mankind that you are mindful of them,
> human beings that you care for them? (Ps 8:3-4)

Like David, Madeleine L'Engle describes how wonder and humility lead to praise when she describes how she feels

---

[55]Norman Wirzba, "The Touch of Humility: An Invitation to Creatureliness," *Modern Theology* 24 (2008): 226.

gazing at the stars: "When I look at the galaxies on a clear night—when I look at the incredible brilliance of creation and think that this is what God is like, then instead of feeling intimidated and diminished by it, I am enlarged—I rejoice that I am part of it."[56]

Have you ever seen the Milky Way? If you live in the United States, chances are you are not able to see the Milky Way without a telescope from your backyard, as more than two-thirds of us live in highly light-polluted regions.[57] Living in environments polluted by artificial light has serious implications for both human health and the health of ecosystems. Light pollution has been shown to have serious effects on humans and other species, primarily due to the physiological effects of altered circadian rhythms. Increased exposure to light at night has been linked to an increased risk of breast cancer among humans and other tumors in wild and domestic animals.[58] Light pollution is also responsible for altered foraging patterns of many wildlife species, changes in predator-prey interactions, and changes in some wild species' abilities to orient during migrations, ultimately threatening global biodiversity.[59] What does it mean that light pollution is diminishing the beauty of creation? The scope of light pollution means that many of us cannot experience our smallness under the stars and understand the scale of creation. Without a view of the stars, how

---

[56]Madeleine L'Engle, *And It Was Good: Reflections on Beginnings (The Genesis Trilogy)* (New York: Crown, 2017), 70.

[57]Malcolm Smith, "Time to Turn Off the Lights," *Nature* 457 (2009): 27.

[58]Ka Yan Lai et al., "Exposure to Light at Night (LAN) and Risk of Breast Cancer: A Systematic Review and Meta-analysis," *Science of the Total Environment* 762 (2021): 143159; Kristen Navara and Randy Nelson, "The Dark Side of Light at Night: Physiological, Epidemiological, and Ecological Consequences," *Journal of Pineal Research* 43 (2007): 215-24.

[59]Franz Hölker et al., "Light Pollution as a Biodiversity Threat," *Trends in Ecology and Evolution* 25 (2010): 681-82.

can we marvel at the idea that the Creator of such a vast and wondrous creation is mindful of us? Wonder helps us understand the scale of creation and how small we are within this scale, and it diminishes the idea that we are powerful. It reorients us to a right relationship with God the Creator and his creation. Wonder helps us know how vast the Creator's love is for us and orients us toward stewardship of creation.

J. R. R. Tolkien describes his character Tom Bombadil as being important in orienting the reader to a proper understanding of humility and power. Tom Bombadil represents "tak[ing] delight in things for themselves without reference to yourself, watching, observing and to some extent knowing" and understanding "the rights and wrongs of power and control" in such a way that the "means of power [become] quite valueless."[60] Bombadil's stories extend the hobbits' journey into times and places that they wouldn't have otherwise experienced, and as they wonder at the places he describes, they begin to understand something more of their place in the story.

> He then told them many remarkable stories. . . . As they listened, they began to understand the lives of the Forest, apart from themselves, indeed to feel themselves as the strangers where all other things were at home. . . . The hobbits sat still before him, enchanted; and it seemed as if, under the spell of his words. . . . Whether the morning and evening of one day or of many days had passed Frodo could not tell. He did not feel either hungry or tired, only filled with wonder.[61]

---

[60]Humphrey Carpenter and Christopher Tolkien, *The Letters of J. R. R. Tolkien* (Boston: Houghton Mifflin, 2000), 179.
[61]Tolkien, *The Fellowship of the Ring*, 129-30.

This time spent wondering seems to put things in perspective for the hobbits.

It is important to understand that Tom Bombadil is only able to tell these stories, thus creating this "wonderful" experience, because he has intimate knowledge of the world in which he lives. In fact, Tolkien explains that Tom Bombadil "embod[ies] pure natural science: the spirit that desires knowledge of other things" but is "entirely unconcerned with 'doing' anything with the knowledge."[62] Dickerson and Evans, in *Ents, Elves, and Eriador*, explain that "Tom Bombadil represents the pursuit and love of selfless knowledge of the created world and its history, independent of any power or advantage that such knowledge might bring to the knower."[63] Thus, we learn from him that we should pursue knowledge of creation, not in order to empower us to subdue creation, but rather to embolden us to wonder at our place among the marvels of creation.

Is it harder to experience wonder today? Not only are we living in a world flooded with artificial lights,[64] we are living in a world flooded with artificial sounds. There are fewer and fewer places on earth where we can experience natural quiet. In the United States, environmental noise, including highway and air-traffic noise, impacts a significant proportion of the population,[65] and we have fewer and fewer places to find

---

[62]Carpenter and Tolkien, *Letters of J. R. R. Tolkien*, 192.

[63]Matthew Dickerson and Jonathan Evans, *Ents, Elves, and Eriador: The Environmental Vision of J. R. R. Tolkien* (Lexington: University Press of Kentucky, 2006), 21.

[64]Hölker et al., "Light Pollution," 681–82.

[65]United States Department of Transportation, "National Transportation Noise Map," 2022, www.bts.gov/geospatial/national-transportation -noise-map; Joan Casey, et al., "Race/Ethnicity, Socioeconomic Status, Residential Segregation, and Spatial Variation in Noise Exposure in the Contiguous United States," *Environmental Health Perspectives* 125 (2017). 1–10.

quiet. Can you believe that human generated sound is double that of background levels in 63 percent of our protected areas?[66] Noises provide important cues to living organisms ranging from those that signal danger and those that indicate the location of a critical resource. Because survival can depend on how an organism responds to environmental noise, many physiological and behavioral systems can be impacted in the presence of excess noise.[67] Studies of the impacts of noise on human health demonstrate that chronic exposure to noise can lead to sleep disturbances, high blood pressure, increased heart rate, anxiety, and mood disorders.[68] Additionally, chronic exposure to noise results in decreased cognitive performance of children and adults.[69] Many additional studies demonstrate that these same effects occur across animal species and suggest that there are serious implications for ecosystem health and biodiversity due to impacts on behaviors like bird songs, habitat selection, and the reproductive success of many species.[70] We read throughout the Scriptures that creation sings! In Isaiah 44:23 we read of mountains and trees bursting into song. In the Psalms we read, "Let the fields

---

[66]Rachel Buxton et al., "Noise Pollution Is Pervasive in U.S. Protected Areas," *Science* 356 (2017): 531-33.

[67]Caitlin Knight and John Swaddle, "How and Why Environmental Noise Impacts Animals: An Integrative, Mechanistic Review," *Ecology Letters* 14 (2011): 1052-61.

[68]Stephen Stansfield and Mark Matheson, "Noise Pollution: Non-auditory Effects on Health," *British Medical Bulletin* 68 (2003): 243-57.

[69]S. Stansfield et al., "Aircraft and Road Traffic Noise and Children's Cognition and Health: A Cross-National Study," *Lancet* 365 (2005): 1942-49; Patricia Tassi et al., "Long Term Exposure to Nocturnal Railway Noise Produces Chronic Signs of Cognitive Deficits and Diurnal Sleepiness," *Journal of Environmental Psychology* 33 (2013): 45-52.

[70]Knight and Swaddle, "Environmental Noise," *Ecology Letters* 14 (2011): 1052-61; Jesse Barber et al., "Anthropogenic Noise Exposure in Protected Natural Areas: Estimating the Scale of Ecological Consequences," *Landscape Ecology* 26 (2011): 1281-95.

be jubilant, and everything in them; let all the trees of the forest sing for joy" (Ps 96:12). In a world polluted with noise, can we hear creation singing?

One of the consequences of living in a noisy, human-dominated world is our failure to recognize (or remember) the natural noises and music of creation. For example, when you experienced the talking beasts of Narnia, were you, like Digory and Polly, enchanted because you did not expect to hear a lion talking? "It was of course the Lion's voice. The children had long felt sure that he could speak: yet it was a lovely and terrible shock when he did."[71] Is it lovely and wondrous to consider that animals can and do communicate outside of Narnia?

I'm sure that it is not a surprise to know that a lion's roar is a means of communication, but have you ever taken the time to consider the complexities of animal communication? Perhaps you thought a lion's roar was a proclamation of impending doom for its prey. Actually, lions roar to maintain social connections, defend territories, and attract mates. Roars are not indiscriminate, as lions refrain from roaring if it will place them at risk.[72] African elephants use infrasonic vocalizations and can recognize individuals from as far away as two kilometers (or 1.24 miles).[73] Giraffes, living in the same habitats as lions and elephants, also seem to communicate long distances with vocalizations described as humming.[74] Mountain gorillas also hum. They hum and sing while they eat,

---

[71]Lewis, *Magician's Nephew*, 117.

[72]John Grinnell and Karen McComb, "Roaring and Social Communication in African Lions: The Limitations Imposed by Listeners," *Animal Behavior* 62 (2001): 93-98.

[73]Karen McComb et al., "Long-Distance Communication of Acoustic Cues to Social Identity in African Elephants," *Animal Behavior* 65 (2003): 317-29.

[74]H. Kasozi and R. A. Montgomery, "How Do Giraffes Locate One Another? A Review of Visual, Auditory, and Olfactory Communication Among Giraffes," *Journal of Zoology* 306 (2018): 139-46.

seemingly to communicate that they do not want to be disturbed.[75] Of course, humans interpret the communications of other species, and we may not actually understand what is truly being communicated; however, taking time to listen to the music of other species can bring us back to wonder and motivate us to stewardship. For example, the discovery in the 1970s that humpback whales communicate with long and beautiful songs has aided in their protection through laws preventing whaling and by encouraging multinational conservation efforts.[76]

J. R. R. Tolkien is well known for his love of trees, and we see trees and forests playing important roles throughout his stories. Anyone who spends time in Middle-earth remembers the ways that trees and forests communicate—whether it is Old Man Willow engulfing the hobbits, Treebeard speaking of his lost tree friends, or the members of the Fellowship listening to the sounds of the forest:

> They rode in silence for a while; but Legolas was ever glancing from side to side, and would often have halted to listen to the sounds of the wood, if Gimli had allowed it. "These are the strangest trees that ever I saw," he said; "and I have seen many an oak grow from acorn to ruinous age. I wish that there were leisure now to walk among them: they have voices, and in time I might come to understand their thought."[77]

---

[75]Eva Maria Luef, Thomas Breuer, and Simone Pika, "Food-associated Calling in Gorillas (*Gorilla g. gorilla*) in the Wild," *PLoS One* 11 (2016): e0144197; Steve Mirskey, "Gorillas Hum and Sing While They Eat to Say, 'Do Not Disturb,'" *Scientific American*, 2016, www.scientificamerican.com/article/gorillas-hum-and-sing-while-they-eat-to-say-do-not-disturb/.

[76]Roger Payne and Scott McVay, "Songs of Humpback Whales," *Science* 173 (1971): 585-97.

[77]J. R. R. Tolkien, *The Two Towers: Being the Second Part of The Lord of the Rings* (London: HarperCollins, 1954), 546.

I wonder if Tolkien understood something about trees that many of us miss? Trees do communicate! I realize that many reject the idea that organisms without nervous systems can be intelligent, but there is mounting evidence that plants interact with each other in intelligent ways—even with plants of different species. If intelligence is defined as a behavioral change in response to an adverse situation and a risk to survival,[78] then many of the observed behaviors of plants—ranging from predation by carnivorous plants to leaf responses and chemical messaging between plants in the presence of an herbivore—could be seen as types of intelligence.[79] The ways that plants communicate work best when plant communities are diverse and undisturbed. For example, fungal associations with roots, called mycorrhizae, are known to increase nutrient absorption from the soil, but they also allow for cooperation between the tree species that share the networks.[80] These fungal networks take years and years to develop and result in a connection between trees. This allows them to communicate via chemicals and share nutrients between plants in adequate light and those that are continually shaded.[81] Trees also communicate with and respond to sounds, both environmental sounds like running water and sounds produced by other plants. Scientists have been able to measure the orienting response of

[78]Paco Calvo et al., "Plants Are Intelligent, Here's How," *Annals of Botany* 125 (2020): 11-28.

[79]Mark Mescher and Consuelo De Moraes, "Role of Plant Sensory Perception in Plant-animal Interactions," *Journal of Experimental Botany* 66 (2015): 425-33.

[80]S. W. Simard et al., "Net Transfer of Carbon Between Ectomycorrhizal Tree Species in the Field," *Nature* 388 (1997): 579-80.

[81]Simard et al., "Net Transfer of Carbon"; Erin Fraser, Victor Lieffers, and Simon Landhäusser, "Carbohydrate Transfer Through Root Grafts to Support Shaded Trees," *Tree Physiology* 26 (2006): 1019-23.

plants to running water, and these same scientists have also discovered that plant roots respond to sounds emitted from other roots—specifically a crackling at 220 hertz.[82] Did you know that 220 hertz is heard by the human ear as an A? I personally find this fascinating because the A at 220 hertz is an octave lower than the pitch that many orchestras use to tune, A-440. When I learned of this connection between the frequency emitted by plants in nature and the frequency used by musicians playing instruments made of wood, it made me wonder. I wonder if we have just stopped listening to the music of creation, so it surprises us when we realize we recognize the tune. If we are lacking evidence that humans can understand the music of creation, there is quite a bit of evidence that other species understand our music. For example, I am amazed that goldfish can learn the difference between genres of music,[83] and that pigeons can distinguish between a major and an augmented chord![84]

Intentionally seeking amazing or beautiful aspects of creation—whether through experience or knowledge—helps us develop the virtue of ecological wonder.[85] Wonderful things happen around us all the time, but we are busy and quickly tune out the chorus of creation. It is easy to tune out the common sights and sounds around us and lose sight of wonder. Have you ever heard the dawn chorus of birds? Do you know which of the flowers in your yard will bloom first? Have you watched a sunrise or sunset recently? We need to

---

[82]Monica Gagliano, Stefano Mancuso, and Daniel Robert, "Towards Understanding Plant Bioacoustics," *Trends in Plant Science* 17 (2012): 323-25.

[83]Ava Chase, "Music Discrimination by Carp (*Cyprinus carpio*)," *Animal Learning and Behavior* 29 (2001): 336-53.

[84]Daniel Brooks and Robert Cook, "Chord Discrimination by Pigeons," *Music Perception: An Interdisciplinary Journal* 27 (2010): 183-96.

[85]Bouma-Prediger, *Earthkeeping and Character*.

rediscover our childlike curiosity, re-engage all of our senses, and rediscover the wonders of creation so that we are motivated toward stewardship. When we are children, there is everything to learn! As adults, we need to reject the idea that we know everything (or enough). When we make the effort to wonder at the natural world, this stance helps us return to a childhood receptivity to learning. As George MacDonald cautions, "To cease to wonder is to fall plumb-down from the childlike to the commonplace—the most undivine of all moods intellectual."[86] C. S. Lewis seemed to be drawn to Christ by his experiences with joy, often in nature. He understood that creation sings God's praises and draws us back to a place where we can experience God: "At best, our faith and reason will tell us that He is adorable, but we shall not have found Him so, not have 'tasted and seen.' Any patch of sunlight in a wood will show you something about the sun which you could never get from reading books on astronomy. These pure and spontaneous pleasures are 'patches of Godlight' in the woods of our experience."[87]

Tolkien's saga of Middle-earth, specifically the fate of Isengard, should be a warning to us when we lose our sense of humility and think we know better than the Creator:

> A strong place and wonderful was Isengard, and long it had been beautiful; and there great lords had dwelt, the wardens of Gondor upon the West, and wise men that watched the stars. But Saruman had slowly shaped it to his shifting purposes, and made it better, as he thought,

---

[86]G. MacDonald, *The Hope of the Gospel* (London: Ward, Lock, Bowden, 1892), 57.

[87]C. S. Lewis, *Letters to Malcolm: Chiefly on Prayer* (London: Geoffrey Bles, 1964), 119-20.

being deceived—for all those arts and subtle devices, for which he forsook his former wisdom, and which fondly he imagined were his own, came but from Mordor; so that what he made was naught.[88]

Are we living in a time of shifting purposes? Saruman embraced the wrong wisdom. His lack of humility broke his connection with creation, and Isengard was no longer beautiful. If we can cultivate a virtue of wonder and humility, we will understand ourselves as a created part of creation, once again hearing and even joining in the chorus of praise to our Creator. When we fail to exercise wonder and delight in the natural world, we open ourselves up to the risk of following Saruman's lead by believing that we have discovered new and better ways to shape creation and utilize its resources. When we approach creation in this way, our actions inevitably lead to depletion of resources, destruction of ecosystems and natural beauty, volatile weather patterns, and the irreversible alteration of our world in ways that harm us all.

In contrast, when we take the time to read landscapes as well as stories, we embrace an important safeguard that reminds us to wonder at the natural world and helps us to better understand our own role as those who are called to humbly serve the Creator in caring for his world. Doing this is not easy. In fact, it requires a radical reorientation of our perspective. As the master of paradox G. K. Chesterton explains,

> The natural thing would be that man should live with the natural things, trees and water and animals, and should, as an exceptional treat, go and look at great buildings

---

[88]Tolkien, *Two Towers*, 555.

and impressive works of art. But for us who live in cities Nature is not natural. Nature is supernatural. Just as monks watched and strove to get a glimpse of heaven, so we watch and strive to get a glimpse of earth.[89]

Thus, the extraordinary thing that Chesterton, along with Lewis and Tolkien, would have us remember is that in learning to rightly view our earth by seeing it through the lens of wonder and humility, we will also receive a glimpse of heaven here on earth in the splendor and beauty of our created world. So, as we strive to live as stewards of our natural world and not as the rulers, we will be richly blessed by receiving the true gift that creation was intended to be from our loving and wise Creator God.

Wonder is a gift, and we do not have to work at all to receive it. Yet it can transform us and move us to love—love for God the Creator, his creation, and also our neighbor. Caspar Henderson says, "Wonder can feel like enough, or like a good point from which to start. It is a state of mind in which we can accept a gift, and apprehend its importance, if not necessarily its meaning. It is a kind of grace."[90] Perhaps this grace helps us to be the people of Christ who embody Jesus' command: "'Love the Lord your God with all your heart and with all your soul and with all your mind' . . . 'Love your neighbor as yourself'" (Mt 22:37-39). To receive this grace, we need to spend more time in creation—"the place where God dwells"—looking, listening, and participating in stewardship. We need to wonder at the fact that we are part

---

[89]G. K. Chesterton, "The Silly Season and Serious Discussion," *The Illustrated London News*, 1905–1907, Lawrence J. Clipper, ed. (Ignatius Press, San Francisco, 1986), 539. Originally published August 31, 1907.
[90]Henderson, *Map of Wonders*, 25.

of this amazing creation and that God has entrusted us with caring for these beautiful places, creatures, and even processes! When we look, we can see ourselves as part of the chorus of creation. We can listen to creation's groanings and praises and also learn about our responsibilities as stewards. Finally, we can communicate and cooperate with each other and ultimately rejoin the chorus.

RESPONSE

Emily Hunter McGowin

I am what you would call a big fan of Dr. Page's work—her teaching, scholarship, and wildlife photography—and I am very fortunate to be able to call her my friend too. I should begin, though, with a personal confession: I don't consider myself an outdoorsy person. I have lived in some of the most beautiful landscapes in the United States, including near the Appalachian Mountains in Virginia, the Rocky Mountains in Colorado, and now in the prairies of Illinois. And yet when I find myself with a rare open day, my first impulse is to curl up on the couch with a blanket and a book. A hike through the woods doesn't typically register in my mind as an option.

I begin with this admission because it's important for me, as a respondent, to acknowledge I am precisely the kind of person who needs to hear and heed Page's message. During my childhood, my life circumstances allowed me plenty of time to explore and play outdoors. I have vivid memories of delving into the woods and creek behind my home in northern Virginia. In my mind's eye, I can still see the various kinds of mushrooms and ferns, tadpoles and frogs, water snakes and garden spiders we found along the way. I spent countless hours back there exploring and make-believing with my friends. But at some point, probably around junior high, I stopped playing outside. And without the influence of wise guides who could usher me back to creation and back to wonder, I will admit that I didn't notice the loss.

Page offers me, and others like me, reason for hope. Drawing on the work of Steven Bouma-Prediger, she has suggested we understand wonder as a virtue. We exhibit the virtue of wonder when we have "the cultivated capability to stand in grateful amazement at what God has made and is remaking."[1] This approach is something of a departure from what we might call classical accounts of wonder,[2] but it remains an intriguing and promising perspective.

The great philosopher and theologian Thomas Aquinas classifies wonder as a species of fear—that is, fear of ignorance. One who wonders, he observes, is at once excited by the novelty encountered and fearful of what is unknown. Yet, Thomas says, wonder is unique because it is the only form of fear that is pleasurable. In his words, "Wonder gives pleasure . . . in so far as it includes the desire of learning the cause, and in so far as the wonderer learns something new."[3] Thus, the one who wonders is compelled to seek the knowledge of things in their causes. So, for Thomas, wonder is a feeling: a special type of fear that is also pleasurable.

---

[1] Steven Bouma-Prediger, *Earthkeeping and Character: Exploring a Christian Ecological Virtue Ethic* (Grand Rapids, MI: Baker Academic, 2020), 43.

[2] See, for instance, the histories provided by Dennis Quinn, *Iris Exiled: A Synoptic History of Wonder* (Lanham, MD: University Press of America, 2002) and Sophia Vasalou, *Wonder: A Grammar* (Albany: State University of New York Press, 2015), or the philosophical account by Mary-Jane Rubenstein, *Strange Wonder: The Closure of Metaphysics and the Opening of Awe* (New York: Columbia University Press, 2010).

[3] Thomas Aquinas, *Summa Theologiae*, trans. Fathers of the English Dominican Province (New York: Benziger Brothers, 1920), I-II, q. 32, a. 8.

To see wonder as a virtue, though, as Page suggests, means it is something that can be cultivated, grown, and lived into Rather than an emotion that simply comes upon you—unexpectedly and outside of your control—wonder as a virtue can be pursued and practiced.

The significance of this distinction might be better understood with reference to other virtues. Thomas Aquinas follows Aristotle in positing four cardinal or primary human virtues: prudence, temperance, justice, and fortitude.[4] These cardinal virtues, he says, are in every human being "inchoatively"; that is, in an imperfect, undeveloped way.[5] But because virtues are essentially *habits*, which Thomas calls "the perfection of power," he says we are able to grow in them. We can become more temperate or courageous, for example, by participating in practices that cause us to demonstrate temperance or courage—choosing moderation in the amount of food we eat, for instance, or choosing to take a hazardous hiking trip every year.[6] Thus, through intentionally cultivating virtuous habits, one can begin to exhibit the virtues without thinking about it; that is, living as though temperance and courage are "second nature" to us.[7]

Whether wonder can be properly understood in the classical sense of virtue outlined above is a question I cannot answer definitively in this brief response. But I can say it is a line of thinking other scholars have suggested too.[8] If wonder

---

[4]The classic modern text on virtue is Alasdair MacIntyre's *After Virtue*, now in its third edition (Notre Dame, IN: University of Notre Dame Press, 2007).

[5]Aquinas, *Summa Theologiae*, I-II, q. 63, a. 1.

[6]Aquinas, *Summa Theologiae*, I-II, q. 55, a. 1.

[7]N. T. Wright offers a very accessible discussion of how Christians can grow in virtue in his book, *After You Believe: Why Christian Character Matters* (New York: HarperCollins, 2010).

[8]It might be more accurate to refer to wonder as a "virtuous emotion" rather than a full-blown virtue as Kristján Kristjánsson does with awe in his book

can be considered a virtue, then it would be very good news for people like me who find themselves regularly bereft of wonder.[9] And there would be a sturdy theological and ethical basis on which to pursue the cultivation of wonder through habits and practices.[10]

Page has given us at least two major ways by which to cultivate the virtue of wonder: interaction with our local ecology and interaction with fictional landscapes. She has demonstrated beautifully for us throughout these chapters what it looks like to attend carefully to the fictional landscapes of Narnia and Middle-earth. As we ponder further what it might look like to intentionally "read" our local landscapes, the Christian tradition offers at least one spiritual discipline we might thoughtfully employ. *Visio divina* (Latin for "divine seeing") is the art of praying with images or other media. Adele Calhoun calls it "a way to pray with the eyes."[11] *Visio divina* is typically used with sacred art—icons, frescoes, or paintings of biblical scenes—but I have also found it works quite well in our own backyards.[12] After all, Psalm 19 says creation declares the glory of God, "pour[ing] forth speech" day

---

Virtuous Emotions (Oxford: Oxford University Press, 2018). Either way, though, he agrees awe (and perhaps wonder too) is morally virtuous.

[9]It would be interesting to bring this conception of wonder as a virtue into conversation with the Catholic tradition of "wonder and awe" as one of the seven gifts of the Holy Spirit, which are received at baptism and then strengthened through confirmation. See the *Catechism of the Catholic Church* (Washington, DC: USCCB, 1994), §1831.

[10]This is something argued by Kristjánsson in his essay, "Scientific Practice, Wonder, and Awe," in *Virtue and the Practice of Science: Multidisciplinary Perspectives*, ed. Celia Deane-Drummond, Thomas A. Stapleford, and Darcia Narvaez (Notre Dame, IN: Center for Theology, Science, and Human Flourishing, 2019). Kristjánsson is convinced, however, that there is an important distinction between wonder and awe, something I will not be pursuing here.

[11]Adele Ahlberg Calhoun, *Spiritual Disciplines Handbook: Practices that Transform Us* (Downers Grove, IL: InterVarsity Press, 2015).

[12]For an introduction to the practice with art, which can serve as an entry-point for the same practice in creation, see Juliet Benner, *Contemplative*

after day (Ps 19:2). What if we prayerfully listened? What would we hear?

The Covid-19 pandemic has given my family and me ample opportunity to become observers of our surrounding wildlife. A well-placed and well-stocked bird feeder has provided countless hours of thoughtful, even prayerful observation of the birds that call our neighborhood home. Beyond this, I can pray with my eyes along with walkers, runners, and cyclists on the Illinois Prairie Path. And I can imagine beginning a hike through nearby Matthiessen State Park with a reading of James Weldon Johnson's poem "The Creation":[13]

> Then God himself stepped down—
> And the sun was on his right hand,
> And the moon was on his left;
> The stars were clustered about his head,
> And the earth was under his feet.
> And God walked, and where he trod
> His footsteps hollowed the valleys out
> And bulged the mountains up.
> Then he stopped and looked and saw
> That the earth was hot and barren.
> So God stepped over to the edge of the world
> And he spat out the seven seas—
> He batted his eyes, and the lightnings flashed—

Vision: A Guide to Christian Art and Prayer (Downers Grove, IL: InterVarsity Press, 2011).

[13]James Weldon Johnson, "The Creation," in *God's Trombones: Seven Negro Sermons in Verse* (New York: Viking, 1927), 17-20. I take theological issue with two lines in Johnson's poem in which he depicts God creating due to God's loneliness. I concur with classical theism, which says God created out of the overflow of the love of the Trinity. God did not need creation, for God lacks nothing, nor could God have experienced loneliness. As a result, creation is sheer gift. Still, I don't think the entirety of such an extraordinary poem should be set aside because of one perceived flaw.

He clapped his hands, and the thunders rolled—
And the waters above the earth came down,
The cooling waters came down.
Then the green grass sprouted,
And the little red flowers blossomed,
The pine tree pointed his finger to the sky,
And the oak spread out his arms,
The lakes cuddled down in the hollows of the ground,
And the rivers ran down to the sea;
And God smiled again,
And the rainbow appeared,
And curled itself around his shoulder.
Then God raised his arm and he waved his hand
Over the sea and over the land,
And he said: *Bring forth! Bring forth!*
And quicker than God could drop his hand,
Fishes and fowls
And beasts and birds
Swam the rivers and the seas,
Roamed the forests and the woods,
And split the air with their wings.
And God said: *That's good!*

How could we not pray, worship, and wonder after reading such a work in the context of God's bountiful creation?

In addition to practices of individual devotion, we must not miss the significance of moral exemplars, or what we might more colloquially call mentors.[14] It is clear that Page was

---

[14]There are innumerable scholarly resources on the significance of moral exemplars in the cultivation of virtue. The body of literature crosses disciplines, too, including philosophy, ethics, literature, psychology, biology, neuroscience, and more. I have been helped in my thinking on the subject by Linda Trinkaus Zagzebski, *Exemplarist Moral Theory* (Oxford: Oxford University

mentored in wonder through her Nanny and Dr. Teska, not to mention the adults who accompanied her on the formative trips to Costa Rica and Ecuador. Those of us who want to be rehabituated to wonder will need similar mentors and moral exemplars. Perhaps Page herself should be among that group. I know she is for me! But I would also point us toward a population we might be tempted to overlook: children. Page encourages us, "We need to be more like children and approach the world with a wide-eyed sense of wonder and endless questions." Yes, indeed. And who better to teach us this than children themselves?

Why would children be exemplars and mentors in wonder? The Christian tradition offers good reason for it. In the Gospels, children are chief among "the least of these" and recipients of the kingdom of God (Mk 10:13-16; Mt 19:13-15; Lk 18:15-17). In a time when children never served as models for adults in anything, Jesus explicitly pointed to them as examples to follow—moral exemplars, if you will, for receiving God's kingdom (Mk 10:15). Jesus seems to think it is precisely in their dependency, humility, and trust that children demonstrate how to embody God's reign on earth.[15]

Given their unique status as recipients of God's kingdom, it does not seem inappropriate to suggest children can serve as teachers, mentors, and moral exemplars in the virtue of wonder. In fact, I have argued elsewhere that children may

Press, 2017). I should note that in scholarly work on this subject, moral exemplars and mentors would be distinguished in important ways. For the sake of this response, though, I am building on what they have in common to bring them together.

[15]Not to mention the fact that Jesus so closely identifies himself with children that to welcome a "little one" is to welcome him (Mk 9:37; Mt 18:1-2, 4-5; Lk 9:46-48). This paragraph is adapted from my essay "Children, Wonder, and the Work of Theology," in *God and Wonder: Theology, Imagination, and the Arts* (Eugene, OR: Cascade, 2022).

have a particular vocation to wonder to which we ought to attend.[16] Perhaps it is time to acknowledge being with children as a vital spiritual practice—not just for so-called "kid people" but for all in the body of Christ.[17] Moreover, it is arguable that children have even more to lose than we do if the people of God cannot regain wonder and rejoin the chorus of creation. Maybe it is time to let them take the lead.

As I bring my remarks to a close, I want to underline Page's call to wonder with a reminder of the telos—or the ultimate goal—to which all of this is moving. Growth in the virtue of wonder, practices of wonder, moral exemplars in wonder— what's it all for? We are creatures of dust ("dustlings" if you will) made to find ourselves in the midst of creation, God's dwelling place (Gen 2:7).[18] Thus, wonder and awe are part of what it means to be human. But many of us have lost sight of this in our highly regimented, largely indoor, screen-focused existence—I myself being the chief of sinners. Certainly, there is good reason to care for creation for its own sake. It is the good gift of a good God, and we are squandering it. But there's a way in which through caring for creation we become more fully human too.

This is not to romanticize nature. Creation, which is both glorious and fallen, remains a dangerous place. We experience wonder and awe in part because of how little control

---

[16]See, again, "Children, Wonder, and the Work of Theology."

[17]Though he doesn't speak of wonder in particular, theologian David Fitch describes being with children as one of the church's vital disciplines for faithful Christian witness in the world. See *Faithful Presence: Seven Disciplines that Shape the Church for Mission* (Downers Grove, IL: InterVarsity Press, 2016), 131-48. Bonnie Miller-McLemore frames care for children as spiritual practice for individual Christians in her book *In the Midst of Chaos: Caring for Children as Spiritual Practice* (San Francisco: Josey-Bass, 2007).

[18]My colleague, Dr. Aubrey Buster, was the first to make me aware of the translation "dustlings" for Genesis 2:7.

we have over the threats of the natural world. But the fact remains: we are *creatures*. Indeed, we are, in the words of Beth Felker Jones, "middle creatures" in whom the spiritual and material realms come together.[19] We, with all of creation, find our ultimate end and purpose in the God-man, Jesus Christ, who unites for all time the transcendent, holy God and the finite, material world. Learning to see ourselves within Christ, and in Christ among the created world—a living bridge between heaven and earth—is central to our vocation as human beings. Learning to, as Page says, "understand ourselves as a created part of creation" is a part of reclaiming our humanity in Christ, the one whose voice rings out over all creation: "I am making everything new!" (Rev 21:5).

---

[19]Beth Felker Jones, *Practicing Christian Doctrine: An Introduction to Thinking and Living Theologically* (Grand Rapids: Baker Academic, 2014), 100-101.

LIST OF CONTRIBUTORS

Kristen Page (PhD, Purdue University) is Ruth Kraft Strohschein Distinguished Chair and Professor of Biology at Wheaton College. Her area of expertise is the transmission dynamics of disease in human-altered landscapes, and her work has appeared in scholarly journals, including *The Journal of Wildlife Management, Animal Conservation, Journal of Wildlife Diseases, Conservation Biology, The Journal of Mammalogy,* and *The Journal of Parasitology.*

Christina Bieber Lake (PhD, Emory University) is the Clyde S. Kilby Professor of English at Wheaton College, where she teaches classes in contemporary American literature and literary theory. She is the author of *Beyond the Story: American Literary Fiction and the Limits of Materialism; The Flourishing Teacher: Vocational Renewal for a Sacred Profession; Prophets of the Posthuman: American Fiction, Biotechnology, and the Ethics of Personhood;* and *The Incarnational Art of Flannery O'Connor.*

Emily Hunter McGowin (PhD, University of Dayton) is assistant professor of theology at Wheaton College as well as a priest and canon theologian in the Anglican Church in North America. She is the author of *Quivering Families: The Quiverfull Movement and Evangelical Theology of the Family.*

Noah Toly (PhD, University of Delaware) is provost of Calvin University. He was previously professor of urban studies and politics and the executive director of the Center for Urban Engagement at Wheaton College. Among other works, he is the author of *The Gardeners' Dirty Hands: Environmental*

Politics and Christian Ethics and Cities of Tomorrow and the City to Come: A Theology of Urban Life, coauthor of Understanding Jacques Ellul, and coeditor of Keeping God's Earth: The Global Environment in Biblical Perspective.

NAME INDEX

SUBJECT INDEX

The Marion E. Wade Center

Founded in 1965, the Marion E. Wade Center of Wheaton College, Illinois, houses a major research collection of writings and related materials by and about seven British authors: Owen Barfield, G. K. Chesterton, C. S. Lewis, George MacDonald, Dorothy L. Sayers, J. R. R. Tolkien, and Charles Williams. The Wade Center collects, preserves, and makes these resources available to researchers and visitors through its reading room, museum displays, educational programming, and publications. All of these endeavors are a tribute to the importance of the literary, historical, and Christian heritage of these writers. Together, these seven authors form a school of thought, as they valued and promoted the life of the mind and the imagination. Through service to those who use its resources and by making known the words of its seven authors, the Wade Center strives to continue their legacy.

THE HANSEN LECTURESHIP SERIES

The Ken and Jean Hansen Lectureship is an annual lecture series named in honor of former Wheaton College trustee Ken Hansen and his wife, Jean, and endowed in their memory by Walter and Darlene Hansen. The series features three lectures per academic year by a Wheaton College faculty member on one or more of the Wade Center authors with responses by fellow faculty members.

Kenneth and Jean (née Hermann) Hansen are remembered for their welcoming home, deep appreciation for the imagination and the writings of the Wade authors, a commitment to serving others, and their strong Christian faith. After graduation from Wheaton College, Ken began working with Marion Wade in his residential cleaning business (later renamed ServiceMaster) in 1947. After Marion's death in 1973, Ken Hansen was instrumental in establishing the Marion E. Wade Collection at Wheaton College in honor of his friend and business colleague.